AF589200

Advance Praise for **IN DEFENSE OF SUNLIGHT**

"The sun provides our civilizations with light, warmth, dinner (via photosynthesis) and of late more and more of the power we use. It also provides our bodies, as this entertaining and informative account makes clear, with what they need. It's the only object in our universe we can't look at directly, but this sideways glance reminds us why we can't live without it."

—Bill McKibben, author of *Here Comes the Sun*

"In this absorbing and fast-paced investigation, Rowan Jacobsen has responsibly reexamined the thinking about sunlight and human health and shown us a plausibly better way. Employing his typically dogged research style, he burns a shoe store's worth of shoe leather getting to the bottom of what is no less than a plot against nature. Avoid this book (and the sun) at your peril."

—Paul Greenberg, author of *Four Fish* and *The Climate Diet*

"Sunlight and its effects on health have become the subject of polarized messaging, but the science is more nuanced than most people realize. *In Defense of Sunlight* offers an engaging look at the evidence and raises important questions about how we think about, talk about, and understand the effects of sun exposure on human health. A fascinating read for anyone interested in health, prevention, and the unintended consequences of oversimplified advice from well-intentioned but siloed medical experts."

—Adewole Adamson, MD, dermatologist and melanoma researcher, University of Texas

"This is absolutely one book everyone needs to read."

—*BookAnon*

"A comprehensive overview of the evidence for the ability of sunlight to promote good health, *In Defense of Sunlight* simplifies scientific jargon and explains why there is harmony between sun exposure and our existence. Jacobsen describes his own awakening to the benefits of sunlight and helps the reader to safely obtain similar rewards."

—Prue Hart, PhD, MS researcher,
Kids Research Institute Australia

"A convincing challenge to the conventional wisdom! We've believed for decades that sun exposure is bad for health, yet the surprising body of science discovered by Rowan Jacobsen and skillfully described in this book tells a very different story."

—Nina Teicholz, author of *The Big Fat Surprise*

"A much overdue wake-up call for the significant public health risks imposed on humans by the modern-built environment and the poorly considered medical advice on sunlight exposure."

—Dr. Geoffrey Guy, founder of the Guy
Foundation and author of *Quantum Biology*

"An illuminating assessment of how solar energy impacts human health."

—*Booklist*, starred review

"The rigorously researched *In Defense of Sunlight* convinced my brain of what my body and soul already suspected: that used judiciously and with respect for its force, the sun is my friend and ally, not an enemy to be shunned."

—Kristi Coulter, author of *Exit Interview*

IN DEFENSE OF SUNLIGHT

THE SURPRISING SCIENCE OF SUN EXPOSURE

ROWAN JACOBSEN

FOREWORD BY

RICHARD WELLER, MD

SCRIBNER

New York Amsterdam/Antwerp London

Toronto Sydney/Melbourne New Delhi

Scribner
An Imprint of Simon & Schuster, LLC
1230 Avenue of the Americas
New York, NY 10020

First Scribner hardcover edition June 2026

Interior design by Kyle Kabel

Manufactured in the United States of America

1 3 5 7 9 10 8 6 4 2

Library of Congress Cataloging-in-Publication Data is available.

ISBN 978-1-6680-9216-3
ISBN 978-1-6680-9218-7 (ebook)

Are not gross bodies and light convertible into one another? And may not bodies receive much of their activity from the particles of light which enter their composition?

—Sir Isaac Newton,
Opticks, 1704

Who shall say of what strange and primitive juices, what fantastic combination of electrons, the true sun-worshiper is made?

—Stuart Chase,
"Confessions of a Sun-Worshiper,"
The Nation, 1929

CONTENTS

PART II
ENLIGHTENMENT

PART III
ALL THE LIGHT WE CAN'T UNSEE

FOREWORD

Richard Weller, MD

I am delighted to have been asked to write a foreword to this fascinating and important book. The dominant message around sunlight today is that it is dangerous stuff that should be avoided, and yet one hundred years ago we had quite the opposite view, with doctors recommending heliotherapy for a whole range of ills. Helped by new scientific technologies, we are again examining our relationship with the sun and rediscovering benefits, some forgotten and some newly revealed. The story of how we got here is intriguing, and I hope that this book will help redress some of the current confusion and imbalance.

Medical science is not a fixed body of knowledge. It continually evolves based on experimentation and the accumulation of evidence. We have moved from the era of leeches and purging to the current world of stem-cell therapies and immunotherapy in a series of small steps based on this gradual acquisition of knowledge.

Sometimes, however, medicine gets diverted onto false trails. As a medical student forty years ago, I spent hours learning the array of operations used to treat stomach ulcers by reducing acid production. To current students this seems bizarre, yet the discovery that gastritis and stomach ulcers are caused by an easily treated infection took fifteen years to percolate through to clinical practice.

Today the medical community is in a similar state of flux regarding sunlight. The growing body of evidence for significant health benefits from sunlight stands in contrast to the prevailing message advising sun avoidance. We need to rethink our approach. Our existing policy on sunlight avoidance stems from advice originally developed for white-skinned Europeans living in sunny countries such as Australia. It is almost certainly inappropriate for people living in lower-sunlight environments or with darker skin types. The medical world is understandably cautious about big changes, for fear that they might be wrong, but also because of the huge institutional investment in the status quo. Syllabi and exam questions have been written, doctors trained, hospitals staffed, guidelines produced, and public-information campaigns run.

Medical silos add to the problem. The adverse effects of sunlight are confined to the skin and thus of interest to dermatologists, yet the benefits—such as reductions in heart disease, cancers, and overall mortality—are not within their field of expertise. Despite this, when it comes to discussing sunlight, internal-medicine experts and oncologists who do have the necessary skills defer to dermatologists.

A rounded view of sunlight's risks and benefits thus falls between these different medical tribes. Added to this is the increasing background noise on social media about photoaging, together with some vociferous influencers saying that sunlight will kill you, and others equally insistent that it will cure you. There are also those who make overstated claims for benefits of vitamin D supplementation. It is no wonder that the whole field is so confusing.

I am sympathetic to my dermatology colleagues who wrestle with the idea of reversing their views on sunlight. It took me decades to work through the process. Like them, I was taught that sunlight is unhealthy and our job as dermatologists is to remind the public of the dangers of sunlight whenever the opportunity presents itself. Among ourselves, however, we would admit that the science was more complex than this. For instance, it has been known for a long time that melanoma occurs more frequently in those who are intermittently exposed to sunlight rather than those who are continually exposed. This was too confusing for public health messaging, so for clarity we put out a crisp and simple "avoid the sun" message.

My views have changed because the data have forced me to. For the last three decades I have been an academic dermatologist, spending half my time seeing patients and half conducting research. There are three main types of research: epidemiology (where you study populations), mechanistic research (where you look at the individual biological processes), and finally clinical trials (where you expose subjects to an intervention and compare their responses to those of control subjects). None of these approaches tells the whole story by itself. They are like separate instruments in an orchestra, individually unimpressive, but powerful together.

I started off doing mechanistic research, studying the effects of nitric oxide on the body's response to ultraviolet light, certain that the oxide must have something to do with skin-cancer development. Only after ten years of this, with results not panning out as I had expected, did I begin to suspect that the mechanisms I'd found might instead be driving beneficial effects of sunlight on blood pressure. So I switched to epidemiology, seeing what the effects of sunlight were on overall health, and finding results completely at odds with what I had been taught in my dermatology training. Now I am back into mechanistic research, with the excitement of a whole world of discoveries to be made.

This is an amazing era in which to be a biomedical researcher. In the UK, we have world-leading population datasets, such as the

UK Biobank, linked to our universal health records. That allows us unprecedented insights into the relationships between behaviors and health. For mechanistic research, whole new technologies have been invented that, together with the data-handling techniques required to understand them, allow us to explore mechanisms and pathways in entirely new ways. Easy communication means that I closely collaborate with colleagues all over the world, complementing and enhancing one another's research.

At home, the University of Edinburgh has been a wonderful place to work, where I am surrounded by talented and generous colleagues and students who have helped steer my research and provided an ideal environment from which to knit together all these different strands. This coming together of resources, data, people, and technologies means, inevitably and necessarily, that we are going to revise and improve our understanding of the world. This may cause discomfort in the dermatology world, but scientific knowledge does not stand still nor should our clinical practice.

This book is thus necessary and timely, steering a path through all the noise, claims, and counterclaims. Beyond that, it is a fascinating read and I am delighted and grateful to see it published. My suggestion is that you take a chair outside, find a sunny spot, turn to the first page, and start to read. I am sure you will enjoy it as much as I have, and you will come away with a much fuller understanding of this important subject.

—Richard Weller, MD
Professor of Dermatology
University of Edinburgh

IN DEFENSE OF SUNLIGHT

INTRODUCTION

HOW I GOT SUN CURIOUS

I finally got right with light in the spring of 2025. It had been a long time coming. By then I'd been writing about the science of sunlight for seven years, but I'd been sun curious for a few years longer than that.

The curiosity stemmed from an ongoing observational study of a single subject: me. I live in Vermont. Gorgeous summers, brutal winters. Over the years, I started to notice what a profound effect the different seasons had on me. Basically, winter sucked. This was not simple SAD (seasonal affective disorder), a common condition in which people feel low during the winter months, and which is believed to be caused by a dearth of light entering the eyes. I didn't feel depressed in winter. I felt as if my cells didn't work.

In summer, my senses were sharper. My thinking was sharper. I ate less but did more, as if nutrients and oxygen were flowing more freely. My skin felt snappier, especially once it got a little color. And I was way happier. Small things seemed deeply pleasing.

And then it went away every winter. This was not due to lack of exercise. I go for a cross-country ski every decent day of winter. I probably get more focused cardio workouts in the snowy months.

But it didn't matter. In winter, brain and body never felt like they were firing on all cylinders.

This wasn't just some inchoate malaise. I could see the difference. In the mirror, I looked a little bit puffy. Despite all the skiing, I'd have a hint of a pouch by late winter. My blood pressure would creep up a few points.

And it wasn't just me. Most people I knew lost their sharpness in winter. Many got depressed. We all commiserated about the short days. Human beings don't hibernate like bears or groundhogs, but our bodies were clearly going into some sort of power-saver mode when the sun withdrew.

It seemed like a no-brainer that it had something to do with light. A bright spring day after a long cloudy spell was like drinking from a well after crawling through a desert. But I never thought much about the mechanism until 2014, when a team of dermatologists at Harvard Medical School confirmed the sun's profound mood-boosting powers. It turns out that when our skin cells are exposed to sunlight, they produce a large molecule called proopiomelanocortin, and, no, that will not be on the final test. Most scientists just call it POMC, pronounced *pom-see.*

POMC is actually the precursor to three essential compounds, and as soon as it's manufactured in the skin, it gets snipped into those three, which have different effects on the body. One stays in the skin and triggers the production of melanin, the dark pigment that makes us tan and protects us from burning, and which is also a powerful antioxidant that excels at mopping up the damage caused by the sun's rays.

The second result of sun exposure is a rise in the production of cortisol, the hormone of activity. It makes your cells fire faster. You get more alert, you burn more energy, and you feel more focused. People think of cortisol as the "stress hormone" and think it's a sign of trouble, but it's only a problem when chronically overproduced in response to too much stress. Its role is to help the body deal with stress—which, in this case, can be anything the world throws at you that requires a response, good or bad. I actually think of cortisol as the "in the zone" hormone. When you are in a tennis match, ready to pounce on a weak serve and fire a winner across court, that's cortisol talking. When you are crushing charades at game night, that's also cortisol. It's the hormone of high performance, of rising to the occasion, and it naturally rises in the morning, but it rises a lot more when we step into the light.

The third compound produced from POMC is an endorphin that heads for the brain and suffuses it with natural opioids, triggering a blissful state similar to what we get from morphine or opium. That one was the focus of the Harvard researchers, who found that blocking the opioids produced withdrawal symptoms. In their paper and accompanying press release, they announced that sunlight is literally addictive.

To me, the discovery had obvious evolutionary implications. What else generates opioids in the brain? Exercise, sex, delicious food, fun social encounters—all the things that keep the species flourishing. Evolution rewards us for these behaviors so we'll keep doing them. In the modern world, these reward systems can be hacked in deleterious ways—addictive junk food, drugs such as heroin that deliver the opioids without the beneficial behavior—but their existence is a sign that people who pursued them prospered and procreated.

So the discovery was somewhat mystifying. Like most people, I'd had it drilled into me since childhood that the sun was bad news, the primary cause of skin cancer, and that the only acceptable amount of its rays that should be allowed to touch my unprotected skin was zero. But if we're wired to seek out sunlight, there must be a reason.

But that wasn't how the Harvard doctors saw it. They acknowledged that the discovery implied some "potential evolutionary benefit . . . that reinforces UV-seeking behavior," but they didn't want to touch that one with a ten-foot parasol. Instead, they focused on how this helped explain why it was so hard to break people of their sun habits. There was even talk of treating recreational tanning as a form of drug abuse.

To me, it seemed obvious that we needed to flip the question around. If sun hitting skin in the morning helped energize us for the day ahead and made us feel great to ensure we'd be right back out there at dawn the next day, might there be other biological benefits we hadn't yet discovered?

More pressingly, what happens to us if sunlight *never* hits skin? By 2014 we were well on our way to becoming an indoor species, a trend that has only accelerated since then. When we did venture outside, we made sure to block the light. Were we depriving ourselves of something essential? Something our basic biology had been built for? And could part of the modern world's malaise be chalked up to sun deficiency?

I began experimenting a bit. Sure enough, a shot of morning light—in my eyes, but also on my skin—led to productive, high-energy days. I also began to notice how much better I felt the instant I stepped outside. Winter, summer, clouds, eyes closed, eyes open, didn't matter, there was always an instantaneous clearing of the cobwebs, a cellular snapping-to. It was as if some nutrient was pouring from the sky.

THAT WAS ALL INTERESTING food for thought, but it stayed a curiosity until 2018, when I began a yearlong Knight Science Journalism Fellowship at MIT. The Knight fellowship is one of those pinch-me positions where you are paid to pursue your intellectual whims into any corner of science they lead. Freed from the pressure to publish, you have the luxury of ignoring the latest hype and drilling all the way down. I went deep on CRISPR and gene editing. I embedded myself with a start-up that was trying to revive the scent of extinct flowers. I stood inside the mirror-lined tokamak, MIT's tiny nuclear reactor, and crunched the numbers on fusion's future. And I read all the research I could find on sun exposure.

My timing was fortunate. The year 2018 was also when a paradigm-shifting study appeared in *Cell*, one of the world's most impactful science journals. The researchers, a neurology team from China, had developed a new technology that allowed them to analyze the makeup of individual neurons. A lot of mind-blowing work came out of the technique, but the paper in *Cell* showed that sunlight hitting skin made neurons work better and improved learning and memory in mice.* The mechanism was a molecule called urocanic acid (UCA), which is abundant in the skin. When hit by sunlight, it mobilizes into the blood and then crosses into the brain, where it's converted into glutamate, a vital neurotransmitter. The extra glutamate reliably increased brain activity and improved mood and cognition. (This is similar to exercise, which also raises glutamate activity in the brain and relieves depression.)

Here was a good reason why the body might be rewarding sun-seeking behavior with an extra dose of endorphins. Not only did sunlight make you happier, it literally fed your brain. This was a revelation, and it was treated as such in the scientific press. "We are all aware that the knee bone is connected to the thigh bone, but

* And not just mice. In 2022, a comprehensive cognitive assessment of 1,838 middle-aged Finns found that those who had been exposed to more sunlight in the preceding few years had significantly better cognitive function than those who lived in darker regions, whose brains were aging years ahead of schedule.

did you know that the brain was connected to the skin?" *Science* magazine, the most prestigious journal in the world, opined in a commentary on the paper. "You'd have to think that beta-endorphin and glutamate are not going to be the end of it (although those two are powerful enough for plenty of effects). One immediately wonders about the well-known effects of sunlight on seasonal depression, among other possibilities."

Two neurologists from Harvard Medical School agreed, penning a commentary in *Cell* titled "Sunlight Brightens Learning and Memory." "Sunlight alters physiology in complex ways," they wrote, "and UCA is likely not the sole mediator of its effects on mood and behavior. . . . This dataset provides a great starting point to delve into other mechanisms responsible for sunlight-induced changes in behavior." And they didn't disguise their position: "Whereas exposure to excessive levels of sunlight is detrimental to our health, moderate exposure can boost our physical and mental state." It would have been nice if they'd strolled across campus to let their colleagues in the dermatology department know that sun avoidance could starve the brain of essential supplies, but everyone stayed in their lane.

In the world of journalism, I seemed to be the only one who was captivated by the news. And the more I dug around, the more astonished I got. The science consistently showed that people who received ample amounts of sun exposure lived longer, healthier lives than people who didn't. Sun deprivation was linked to heart attack, stroke, diabetes, dementia, depression, Parkinson's, myopia, respiratory infections, certain cancers, and autoimmune conditions such as multiple sclerosis and inflammatory bowel disease.

That evidence had been showing up in various observational studies for decades, but it had been growing stronger as the tools got better. For instance, one massive UK study attached light-sensing watches to the wrists of 89,000 volunteers to document their weekly light exposure, then followed the group for eight years to see who died and who didn't.

Who died? Those who didn't get enough daylight. People exposed to the most light were 34 percent less likely to expire than people who got less daylight than average.

Maybe that was just because they were exercising more? That's the right question to ask, but these are the things epidemiologists wake up asking themselves. We know that lots of lifestyle factors affect health, including smoking, exercise, diet, age, socioeconomic status, and so on. So to be sure they are comparing apples to apples, all modern observational studies account for well-known confounders. In the case of the UK light study, even after adjusting for everything from exercise and diet to smoking and sleep duration, the sun getters were still 17 percent less likely to die. As the authors put it in their paper, "Personal day light exposure appears to be an independent predictor of mortality risk."

Why? That's what this book is about. The short answer is that the human body evolved in an environment of abundant sunlight, and we are designed to make use of that solar energy in various ingenious ways—ways that we are finally starting to understand.

If you're wondering why you've never heard about any of this, well, join the club. After years of following the research and wondering why it wasn't making its way into the public sphere, I've concluded that part of the problem is simply that the sun has no lobby. There are a lot of people whose job it is to tell you about the risks of too much sunlight, and a lot of companies hoping to sell you some product to protect yourself from it. In contrast, it's nobody's job to tell you about the good stuff.

So I've made it my job. This book is an attempt to weave together the threads of dozens of different scientific investigations over many decades, to tell the full story of light and how our bodies use it.

I think it's an important story to tell because we can all benefit from it. The medicine, after all, is right outside the door. Each extra hour spent outdoors lowers the rate of cancer, diabetes, obesity, and depression.

It's certainly worked for me. A simple commitment to a little more light each day has profoundly boosted my well-being, but it's also reenchanted my experience of everyday life by changing the way I think.

I USED TO THINK of light as ephemeral, a mere backdrop against which the drama of living matter unfolds, but light is just as physical as any atom. It's pure energy, bundled into tiny packets called photons that zip out of the sun at 300 million meters per second and find their way to Earth eight minutes later, where they interact with everything they touch, including you.

The photons that emerge from the sun have a huge range of energy levels. The more energy they have, the faster the frequency at which they oscillate, and the shorter their wavelength as they travel. (For a deeper exploration of the physics of light, see the color insert "A Light Primer.")

When a photon of light meets a molecule that is oscillating at a similar frequency, it will get absorbed by that molecule, energizing it. That happens billions of times per second in our bodies. It's how we see. We have receptors in our eyes tuned to absorb different frequencies of light, and we see these different frequencies as colors.

But photons continuously bombard our entire bodies, not just our eyes, merging with molecules of matching frequencies and making them dance. We depend on that dance to set our internal clock, to make essential compounds, to get a sense of the world around us, and to do a dazzling array of vital tasks, as I'll explain.

Because different colors of light have different frequencies, they have different effects on the body. Blue, for example, energizes us. Green soothes us. But all the visible frequencies, from red to violet, comprise less than half of the light coming from the sun. Those are just the photons we can detect with our eyes. About half the solar spectrum is infrared light, which has lower energy than visible light. We can't see it, but it's pouring into our body whenever we're outside (we feel it as heat), and it actually penetrates far more deeply than the other colors. Scientists now suspect that these gentlest of photons may heal damage at the subcellular level, and there are even hints that they can help stave off aging.

On the other end of the spectrum, beyond violet and also invisible to us, is ultraviolet light, which accounts for just a few percent of the solar spectrum but which presents some of the biggest challenges and opportunities. An ultraviolet (UV) photon has so much energy that it can actually scramble the atomic bonds that hold molecules together. That's why UV can be damaging. Some UV photons have just the right frequency to be absorbed by DNA molecules themselves. Like a direct hit from a photon torpedo, that energy overload can disrupt the molecule. If not fixed, that can lead to mutations and cancer.

Fortunately, because UV has such short wavelengths, it penetrates only a few hundredths of a millimeter into the skin before hitting a molecule, so any damage is limited to the surface. And, of course, life has been dealing with UV for 4 billion years. Organisms have invented complex systems of damage control, as well as ways of harnessing the energy of UV to produce valuable molecules such as vitamin D and nitric oxide.

Overall, the ledger leans strongly toward the upside of sun exposure, as we know from the excellent mental and physical health of people who get more of it. Full-spectrum sunlight produces a pharmacopoeia of beneficial compounds in the body that substantially reduce the risk of every lifestyle disease you can think of. It also provides the body with essential cues that help us calibrate our biology to the changing days and seasons. It's an elegant system that's been fine-tuned over billions years of existence on a sun-drenched planet.

But it's also one that's been thrown into disarray in recent times with our retreat from the outdoor world. Most Americans spend less than an hour per day outside. One in five virtually never sets foot beyond the built environment. That leaves us trapped in what chronobiologists call biological darkness—weak and weird artificial lighting that is so different from full-spectrum sunlight that our cells can't make use of it.

Cut off from its ancient lodestar, the body loses its way. The brain gets trapped in a never-ending pit of seasonal affective disorder. The ramifications are so alarming that fifteen scientists with expertise in these fields recently published a paper titled "Insufficient Sun Exposure Has Become a Real Public Health Problem," in which they estimated that 820,000 deaths per year in Europe and the United States stem from sun deficiency. We need to find a better balance, and the first step toward that balance is the one that leads outside.

A DECADE AGO, most experts thought they knew why sun getters were so healthy: vitamin D. A century earlier, scientists had discovered that sunlight hitting skin produced D, a hormone that regulates calcium in the body and is essential for strong bones and a number of basic biological processes. The effect is so predictable that D is actually a great measure of how much sun exposure a person has received: the more sun you get, the more D in your blood.

In the second half of the twentieth century, doctors began to notice that people with higher blood levels of vitamin D also had lower rates of diabetes, obesity, osteoporosis, heart attack, stroke, depression, cognitive impairment, autoimmune diseases, colorectal cancer, respiratory infections, you name it—130 different health outcomes in all. It seemed that the doctors had stumbled upon a new wonder drug, and that raising everyone's D levels could result in massive health improvements.

The easiest way to do that would have been through sun exposure, but the public was already being urged to avoid sun exposure because of skin cancer. So the authorities turned to vitamin D pills, which quickly became the most prescribed supplement in the world. Clinical trials testing its effectiveness hadn't yet been completed, but D looked so promising that doctors saw no reason to wait for verification.

In hindsight, it was naive to think that a pill could replace an intricate biological system, but science has a long track record of such reductionist mistakes. The results of the trials started to come in right around the time I was doing my research at MIT, and they shocked the medical world. Supplementation did indeed raise the levels of D in people's blood, but it failed to improve a single condition. The results were consistent across hundreds of trials, a rare consensus. Top publications such as *The New England Journal of Medicine* called for doctors to stop prescribing D. Top magazines such as *Scientific American* ran features debunking it. Though it is still widely prescribed, the supplement is now considered useless for anyone except those who have a severe deficiency of the vitamin.

But that left a mystery. If D wasn't the reason people who got more sun exposure were so much healthier, what was?

By then I was hooked. It turned out many scientists were asking the same question. Immunologists were finding that sun deprivation during childhood led to higher rates of autoimmune diseases. Cardiologists were noting lower blood pressure and less diabetes in those who got a daily dose of sunshine. Endocrinologists were realizing how many hormones and neurotransmitters were produced when sunlight hit skin. Chronobiologists were seeing accelerated aging in night-shift workers and astronauts. Neurologists were discovering the surprising ability of certain wavelengths to reduce pain and enhance wound healing. And psychologists were learning that sunlight's ability to keep the mind on an even keel went way beyond what could be achieved with a SAD lamp. The skin was turning out to be the organ that set the tone for the functioning of the entire body, and we'd unwittingly been sabotaging it for decades.

In 2019 I finagled an invite to the first-ever conference on sunlight and health in Washington, DC. I was amazed to step into an alternative universe of scientists who had often quietly been studying the health effects of light for decades. Like many scientists, they were more focused on doing research than spreading the word, so few people had heard about their work. But they had no objections to genuine interest in their discoveries, and they graciously fielded my endless questions.

One of them was Richard Weller, a professor of dermatology at the University of Edinburgh, who has been one of the few in his field not to shy from public discourse. The first time I interviewed Weller, I made the mistake of characterizing the idea that sunlight could be beneficial as counterintuitive. "It's entirely intuitive," he responded. "*Homo sapiens* have been around for two hundred thousand years. Until the industrial revolution, we lived outside. How did we get through the Neolithic era without sunscreen? Actually, perfectly well. What's counterintuitive is that dermatologists run around saying, 'Don't go outside, you might die!'"

But in truth Weller and the other experts I've spoken with over the years do understand how we came to such an extreme position on sun exposure. They see it as an unfortunate consequence of the increasing specialization of the sciences. The institutions crafting the policies and messaging on sun protection have one job to do: reduce the risk of skin cancer. That's a laudable goal. There are 5.5 million cases of skin cancer each year in the United States—more than for all other types of cancer combined. While almost all these cases are minor, it would still be great to drive that incidence as low as possible. And the most obvious way to do that is to discourage anyone who might be tempted to stick a toe in the sun.

By that metric, the powers that be have done an extraordinary job. The media is flooded with stories about the dangers of skin cancer and the need to wear sunscreen. Hats off to the agencies responsible. In terms of getting the message out, nobody has done it better.

But over the years, as rates of skin cancer continued to rise despite massive sun-protection efforts, what started as a campaign to prevent overexposure morphed into a kind of fundamentalism. Protect yourself when you're at the beach. No, cover yourself anytime you'll be out and about. Actually, wear sunscreen like a second skin.

"Sun protection is necessary every day, regardless of the weather or time of year," the Skin Cancer Foundation tells us. It wants us to use sunscreen every day and reapply every two hours. That message has dutifully been conveyed by the mainstream media as well, generally in terms similar to what *Self* magazine suggests in its "Ask a Derm" series: "It is *always* a good idea to wear sunscreen daily—whether it's sunny or cloudy, summer or winter, warm or cold, or you're planning on being outside or staying indoors."*

But one result of such a relentless campaign is that the dangers of sunlight loom larger in the public mind than they do in the actual

* Even top researchers have no clue where such an idea came from. "This is totally the wrong advice," the dermatologist Brian Diffey, who designed the UK rating system for sunscreens and has published hundreds of papers on their efficacy, told *Vogue* in 2025. "It's not a good habit. It's a bad habit. You shouldn't do it."

data. I'd always assumed that skin cancer is one of the leading killers, but it isn't. Not even close. Nearly 99 percent of skin cancer cases are basal cell carcinomas (BCCs) and squamous cell carcinomas (SCCs). Those are certainly no joke. They can be unpleasant and unsightly, they can leave scarring, and if allowed to get large they can interfere with sensitive areas such as eyelids, but they rarely spread beyond the skin or result in serious health issues. They just need to be removed.

"When I diagnose a basal cell carcinoma in a patient," Weller told me, "the first thing I say is congratulations, because you're walking out of my office with a longer life expectancy than when you walked in." It's not that BCCs are good for you; it's just that the people who get them tend to be healthy outdoorsy types making the most of their golden years, and the BCCs don't even slow them down. Melanoma is far more dangerous, but it's also uncommon, accounting for less than 2 percent of skin cancers, and its associations with sun exposure are less straightforward.

Globally, skin cancer doesn't even make the list of the forty deadliest killers. It's responsible for about 120,000 deaths per year, which pales in comparison to the 20 million lives lost each year to cardiovascular disease or the 10 million lost to other forms of cancer. It would be great if we could wave our wand and make skin cancer go away, but recommending a practice that lowers rates of skin cancer while raising rates of other, more problematic diseases makes no sense.

And it's especially poor advice for people with darker skin, who almost never get sun-induced skin cancer and yet suffer higher rates of many of the diseases that seem to be alleviated by sunlight. So it's one thing to debate the risks and benefits of sun exposure for the small portion of the global population that has fair skin, but for everyone else there's simply no question: Moderate sun exposure provides tremendous benefits, and one-size-fits-all recommendations are misguided.

In truth, authorities know that their messaging on sun exposure is simplistic to the point of inaccuracy, but they fear that if the message is more complicated than "just say no," some people will get it wrong. So the goal has been to instill sun avoidance and sunscreen use as an automatic habit, no thought required.

I do understand the need to dumb down the messaging for the modern media environment. But this isn't a public service announcement or a TikTok reel. It's a *book*, an arcane technology designed specifically for people who like to grapple with big, complex, nuanced subjects and make their own decisions. If you've made it even this far, you are in the right place.

IF YOU'RE SURPRISED THAT you haven't heard more about the benefits of sun exposure from your health-care providers, don't be. It takes medical and scientific institutions much longer to incorporate new discoveries into their recommendations than you'd think. Information does not flow freely. There is friction at every stage: institutional inertia, health-care providers still going by what they learned in medical school, health columnists hesitant to undo years of advice, journal editors and reviewers who don't want to see their life's work undermined. Scientific revolutions generally proceed, as the quantum physicist Max Planck famously put it, one funeral at a time.

And if information is slow to update within a field, it is even slower to cross over to an unrelated field, especially as modern science has become a world of silos and microspecialties. Most scientists spend their careers learning everything there is to know about a single piece of the megapuzzle. They're unlikely to learn about the latest developments in cardiology or immunology unless they stumble into the wrong lecture hall by accident.

"Our medical tribalism and the disappearance of the old-fashioned generalist has put us in this situation," Richard Weller told me. "Neurologists and cardiologists faced with the barn-door-obvious benefits of sunlight for their diseases defer to dermatologists preaching about the lethality of sunshine."

Weller has the kind of inquisitive mind that was de rigueur for scientists of a previous generation, which has led him to investigate the fundamental implications of the human relationship with sunlight in a way that's rare in today's world. But that makes him a man without a country. "There's a bunch of us around the world who realize sunlight is good for you, and there's no place for us," he said. "I publish in kidney journals, neurology journals, epidemiology journals, cardiovascular journals, hypertension journals. I don't publish much in dermatology journals because they're just not interested."*

But I was interested. In a sense, science writing is the last holdout of the old-fashioned generalist, the person whose job it is to stay alert for interesting discoveries that tell us something new and useful about the world, to connect the dots across a constellation of data points until, one hopes, an interesting picture emerges. In the case of light and health, that picture gets quite complicated, but I've found it to be of interest to almost everyone—because almost everyone is a little bit sun curious.

THE TITLE OF THIS BOOK is a nod to *In Defense of Food*, Michael Pollan's 2008 "eater's manifesto," which called for a return to nutritional common sense. Pollan exposed the bad science driving

* The exception is the *Journal of Investigative Dermatology*, which has been chronicling the new science of sun exposure for decades, and which you'll find frequently cited throughout the notes section of this book.

"nutritionism," the era's nutrient-by-nutrient approach to eating, and urged us to avoid eating anything our great-grandmothers wouldn't have recognized as food, offering his now-famous seven-word prescription: "Eat food. Not too much. Mostly plants."

The title of Pollan's book was intentionally puzzling. Why would someone need to defend something as fundamental as food? The answer was that nutritional pseudoscience and a food industry ready to capitalize on it had obscured the basic wisdom that had served humanity well for thousands of years, scaring us into the arms of "edible food-like substances" with dubious health claims.

If you ever think the experts can't all get it wrong together, just think of all the people you know who eat egg-white omelets or avoid eggs altogether because they have cholesterol in them. The idea that eating cholesterol would raise the levels of it in your blood was junk science from the 1930s that had been debunked by the 1950s, yet seventy years later some doctors still haven't gotten the memo.

That wasn't the only bit of dismal nutritionism that wouldn't die. Based on little more than the fact that a gram of fat has nine calories while a gram of sugar has four, experts told us to avoid fat in favor of ultra-processed low-fat products that were often sky-high in sugar. It was terrible advice, resulting in a surge in obesity and disease, and it seems laughable to us today, but it was dogma in all the top institutions and agencies, and heretics who questioned it had their careers burned at the stake. Then the inquisitors blew it again, replacing butter with trans-fat-laden margarine that turned out to be one of the deadliest foods ever foisted on the public.

By 2008, better science was providing a wealth of evidence showing that traditional eating had been healthier all along. Yet there was still a pressing need for a book such as Pollan's. An empire of careers had been built on nutritionism, and the loudest voices of all were the companies with goods and services to sell. Real food needed defending.

And now, it seems, so does another fundamental nutrient, a kind of whole food we evolved with and abandon at our peril.

If we are to abandon it, the burden of proof lies with those urging such a radical departure from evolutionary norms. That's the lesson we keep learning with all the diseases of civilization. Our bodies evolved for a particular environment and lifestyle, and the most successful health interventions have been the ones that put us closer to those preindustrial norms: eating more leafy greens, for example, or getting more exercise, or avoiding cigarette smoke and microplastics. Those all pass the evolutionary sniff test. But if you want me to ingest some new substance or stop some behavior that my ancestors happily practiced for millennia, well, I'm open to the idea, but you'd better show me some extraordinary evidence first. And so far, the experts urging solar abstinence have not cleared that bar.

One of the most thoughtful observers on matters of the skin is the Penn State anthropologist Nina Jablonski, author of *Skin: A Natural History.* Like me, she sees similarities between current calls for solar abstinence and the misguided diet advice of the 1990s. "A lot of people just got very irritated and said, 'I'm not going to pay attention to this stuff,'" she says. "And that's what we face now with these punitive, one-size-fits-all approaches. It just goes against human intelligence and human emotions, and people will reject it."

Jablonski told me that a few years ago, and clearly she was ahead of the curve. Most people still dutifully duck the sun without questioning the logic. But there are signs that things are changing at last, that people are ready for a more grown-up conversation about sun exposure. Jablonski sees the field of nutrition as a promising example of our ability to engage with new science when it impacts our personal health. "In the past twenty years, people have come to grips with fairly sophisticated information about diet. My feeling is that they can get their heads around some of the subtleties of sun exposure, too."

Mine, too. So I'll offer my own seven words of advice: *Get sun. Not too much. Go outside.*

That last bit is important. It's difficult to untangle the multitude of benefits people receive when they're outside. In addition to sunlight, there's exercise, fresh air, mental decompression, improved circadian rhythms, probably other things we don't yet understand. It almost certainly works best as a package.

In contrast, we have compelling evidence that indoor life is deeply unhealthy. So whatever the fine points, the prescription is the same: *Go outside.* There's no need to overthink it. Wear a hat, wear sunscreen, do what you gotta do, just get out there.

That's right, don't throw away your sunscreen. Arguments over the benefits of sun exposure often devolve into fights over sunscreen, but the two can be complementary. If sunscreens get people outside and allow them to reap the benefits of wavelengths of light other than UV, that's great. Most sunscreens also allow a smidge of UV to sneak through, and often a smidge is all you need.

That's another important consideration. You are not a cactus. Even the researchers most enthusiastic about the sun's salutary effects recommend small doses, depending on your individual situation. If you have fairly pale skin and live in a sunny place, you need just a few minutes a day at most, and you probably get that without even trying. Conversely, if you have dark skin and live in a low-sun environment, you might benefit from a lot more. In between those extremes lies a huge gradient that is further affected by many other variables, including health, age, gender, diet, season, and more. The important thing to remember is that once you understand a few basic principles, it's easy to get most of the benefits of natural sunlight without incurring any significant risks.

The rationale behind all of this will be explored in the pages to come, as well as in the color insert, which elucidates the physics of light. Understanding that science can help you see the world in an entirely new and beautiful way, and it's a key to all the new thinking on light and health. But it's also the one part of the book with lots of hardcore science, so if physics isn't your thing, feel free to skip it. You can always come back to it as needed.

From there, part 1 tells the fascinating story of our changing attitudes toward the sun, from pillar of health to feared carcinogen. You'll learn how people with different skin colors can handle vastly different amounts, and how it can go wrong when sun and skin are a bad match.

Part 2 chronicles the recent discoveries in a range of fields that have revolutionized our understanding of light and health, and it introduces some of the researchers who were curious enough to follow the science where it led, even if it caused them professional hardships.

Part 3 explores the huge role light plays in sleep and energy metabolism. I'm the lab rat in that one, relaying my adventures and misadventures on the circadian path in search of a lightscape that is truly healthy. By the end, you'll understand all the factors I incorporated into my personal heliotherapy program and will be ready to begin your own.

But beyond the practical side of this book, I'm hoping it has a spiritual side, too. That, I think, is the most precious thing we've lost as we've come to see the sun as an enemy. It casts a shadow across our entire experience of the outdoors, of nature, of the world we were born into. It can trick us into seeking the illusory safety of indoor environments with unhealthy lighting.

That trend has been underway since the industrial revolution, in what amounts to a vast experiment to find out what happens when a species that evolved to live in the elements trades up for an indoor existence centered increasingly on the wrong kinds of rays. It will come as no surprise that the results are concerning, especially in younger generations who know nothing else. We are forgetting the heavens.

What's most ironic is that, in a sense, we *are* the heavens. The difference between you and a doorstop is the energy that animates your body. Without that, no movement, no consciousness, no you. That energy comes from sunlight. Most of it fell to Earth, embedded itself in a plant, and then found its way to you. But some of it found you directly. Either way, we are lumps of mud animated by light and having a conscious experience of the universe, and we depend on constant infusions of light to keep the fun going. The good news is that—as of this writing, at least—sunlight is still free, legal, and widely available. Get it while you can.

PART I

THE APE THAT FELL IN LOVE WITH THE SUN

CHAPTER 1

SOL MATES

Two Million Years Under African Skies

I've always had a strong hunter-gatherer streak. I grew up hunting golf balls on a little course in New Hampshire. I vividly remember the thrill of padding silently through the woods, eyes tuned to any hint of white, and then suddenly spotting my quarry nestled against a spruce tree. Now I hunt chanterelles. Different color, same thrill. For work, I hunt stories. It's no different. You roam the landscape, searching for overlooked nuggets.

But we've all got a little hunter-gatherer in us. A few thousand years of agriculture and houses has hardly altered our bodies from their prehistoric plans. They are not designed for swinging scythes or tapping keyboards. Ask me how I know.

The reality that we are essentially hunter-gatherers shoehorned into a twenty-first-century existence shows up in many biological mismatches. What food is good for us? The stuff we got before industrial agriculture: roots, leaves, fruit, a little antelope. What's bad for us? Processed starch.

What else is good for us? Walking, running, fresh air, fresh water, hanging out with friends and family.

And then there's light. There's a distinct light environment we're adapted to, and it's not the one you find in a man cave. We began adapting to it a million years ago, maybe two, somewhere in Africa's Great Rift Valley. Around that time, we underwent a series of rapid-fire evolutionary changes that transformed us from just another promising primate into something the world had never seen. We left our swinging ways behind and decided to try our luck on the ground, with feet that adapted to walking and running. Our expanding brain turned us into skilled hunters and fire makers.

But another biological breakthrough around the same time also deserves attention. We are the ape that stepped out of the forest, dropped its fur coat, and began working on its full-body tan.

At first glance the whole thing seems counterintuitive. What's a self-respecting primate doing trading the safety of the trees for the dangers of the savanna, and compounding the error by going naked under a fierce African sun? None of the other mammals of the region seem to have embraced this strange evolutionary path. But we were on a whole different trajectory.

Anthropologists have spilled a lot of ink puzzling over the odd fact that we lost our fur at the same time that we left the forest for the sunny savanna, where we'd seemingly have needed the protection all the more. One interesting piece of evidence is that it happened in lockstep with the development of a ridiculous number of sweat glands. Compared to chimps, humans have ten times the density of sweat glands. Getting rid of fur allowed us to dump heat via sweat like no other creature, a unique cooling system that is one of our defining features. We should have been named *Homo perspirus.*

That cooling ability allowed us to raise the metabolic rates of our bodies and brains to an extreme level for an animal our size. Other residents of the Serengeti can outsprint us, but they need a lot more downtime to literally chill out. Chimps, for instance, spend all of ninety minutes per day in active pursuits. They spend the other ten and a half hours of their waking life grooming, idly munching on leaves, and generally loafing around. Then they sleep for twelve hours. In contrast, we burn the grassland at both ends, always on the make, exploring, scouting, inventing, experimenting, outsmarting. And working up a good lather the whole time.

That still left the issue of the fierce sun. Originally, like most primates, we had fairly pale skin that was susceptible to burning. Part of the solution was to coat ourselves in a protective layer of microbes. All the sweat being produced by those abundant sweat glands is excellent food for a menagerie of friendly bacteria that colonize our skin. Oil, which we also produce more abundantly than other primates, is another microbe favorite.

A typical square centimeter of human skin is colonized by 1 million bacteria, along with millions of other assorted microbes. In addition to protecting us from infection by pathogenic organisms, this skin microbiome forms a kind of living sunscreen composed of species that are adapted to lots of ultraviolet exposure. They readily absorb UV and use it for their own needs, and they proliferate in sunlight. The skin of lifeguards at the beginning of a season has few of these beneficial bugs, but by the end of summer it's packed.*

The radiation that got past these microbial sentries was handled by our other great innovation to life on the savanna. We evolved dark skin rich in melanin, a pigment that is incredibly good at absorbing ultraviolet light before it can cause trouble. For good measure, our skin also responds to sunlight by thickening, adding layers of protective cells that are constantly sloughed off. "For eons," says the biological anthropologist Nina Jablonski, "skin stood up to the sun."

* Intriguingly, some of these microbes actually produce compounds that can directly kill cancer cells without harming normal ones.

At the same time, we put that free energy to work, synthesizing a pharmacopoeia of useful compounds in our solar-powered cutaneous factory. By wedding skin and metabolism to the sun far more intensely than any primate before us, we became fully human, child of the light and master of the savanna. That sunlit environment of grasslands and woodland edges is what we're designed for. For evidence of this, look no further than the sun-induced endorphins I mentioned in the introduction. Sun seeking is built right into the specs.

BACK THEN, WE ALL had dark skin, and we stayed that way for at least a million years. We spread buck naked across the continent and eventually the planet, pushing into colder regions about seventy thousand years ago with the help of clothing and fire.

But things were different in the high latitudes. Due to the tilt of Earth, the farther you get from the equator, the lower the angle of the sun in the sky. And the lower the sun, the more atmosphere its rays must pass through to reach the ground. More atmosphere, more humidity, more cloud cover—it all reduces the number of photons that reach the ground. So while the Great Rift Valley might receive an average of 2,000 kilowatt hours of solar energy per square meter in a year, Sweden gets just 1,000. The problem is exacerbated by the need to wear clothing in colder regions. When it's freezing, you don't expose a lot of skin.

As populations migrated toward these low-sun areas, they kept making less of that protective melanin and their skin got thinner and paler to let more sun in. This happened multiple times in prehistory—in Scandinavia, on the steppes, in East Asia, even in Neanderthals—and it happened through different genetic combinations, which is a good sign that it wasn't random. In lower-light environments, even when people were spending their whole days outside, less sun blocking was advantageous. Light skin is a sign of how important sunlight was to survival—precious enough to warrant some significant biological tinkering.

Because of these adaptations, the global distribution of human skin tones tracks closely with annual solar radiation. Experts divide that spectrum into six different skin types, a system known as the Fitzpatrick scale, which is far from perfect, but still useful shorthand. An alabaster-skinned redhead would be a I on the Fitzpatrick scale, while a person with typical Northern European ancestry would be a II. People with Mediterranean and Near Eastern backgrounds are usually types III and IV, those from Asia and Latin America types IV and V, and people with African and South Indian ancestry are mostly types V and VI.

The amount of melanin in skin ranges from virtually none in type I to 25 percent in type IV and a whopping 60 percent in type VI, which is enough to provide the equivalent of SPF 60 in the base of the epidermis, the outermost layer of skin.

But even people with very little melanin can safely benefit from the sun. In 2016 a team of dermatologists at the University of Manchester wanted to find out if pale-skinned people could raise their vitamin D levels using small amounts of sun exposure without raising their risk of skin cancer. They exposed volunteers to the equivalent of fifteen minutes of UK summer sunshine three times a week for six weeks. That small dose was enough to give them a significant boost in vitamin D.

Then the dermatologists took biopsies from the volunteers' skin and checked it for damage.* The light dose of sunshine caused

* Scientists have developed a clever way of doing that. When UV photons collide with DNA, their energy causes parts of the molecule to stick to each other, distorting the shape and potentially leading to mutations in the genetic code. The immune system tags these distortions with antibodies, marking them for future repairs, and scientists can determine how much damage has occurred by counting the antibody tags.

modest amounts of DNA damage. It always does in pale-skinned people, which is why the authorities say you should avoid all direct exposure. But what came next is important. As the study continued, there was no accumulation of damage. It was all fixed shortly after occurring. We all have damage-repair enzymes in our bodies that spend every day fixing broken DNA.* They aren't perfect, and they can be overwhelmed by too much damage, but the more practice they get, the better they become, and the damage caused by fifteen minutes of summer sunshine turned out to be trivial, while the benefits were significant.

That study was published in the *British Journal of Dermatology*, along with an accompanying editorial by Michael Holick, the world's foremost expert on vitamin D. "You can have your cake and eat it too," Holick wrote, undoubtedly annoying his peers. The results of the study, he suggested, "should provide healthcare regulators, especially those who have advocated sun abstinence because of their concern for increased risk for skin cancer, with a new perspective for how the sun and our skin work in concert to take advantage of the many beneficial effects of sensible sun exposure while minimizing risk for skin cancer."

So "sensible" sun exposure, as Holick calls it, seems low risk and high reward. But what about those light-skinned people thousands of years ago who spent all day outside without a drop of Coppertone within reach? Wouldn't they have been eternally sunburned?

Probably not, says Nina Jablonski, the Penn State anthropologist. "Even for people with lightly pigmented skin, painful sunburns would have been exceedingly rare because there was never a sudden shock of strong sun exposure. Rather, as the sun strengthened during spring, the top layer of their skin would have gotten gradually thicker over weeks and months of sun exposure."

* That discovery earned a Nobel Prize in 2015.

Skin responds to even small amounts of sunlight by thickening and darkening in preparation for more to come. It gets into trouble when it's caught unprepared—something much more likely now, when we go straight from the winter office to the Caribbean sand. In other words, it's not a design problem, it's a lifestyle problem. As the dermatologist Richard Weller put it in a paper for the American Heart Association, "Since life on earth developed under the influence of solar radiation, it is difficult to imagine that human physiology should be unable to cope with UV radiation; more likely, modern man is 'out of sync' with this important environmental stressor."

So, yes, light skin is much more sensitive to sunlight and prone to skin cancer than dark skin, but all of us are built for some amount of sunlight.

For nearly all of human history, this was not a controversial statement. Longing for light seems to be universal. Most cultures have implicitly understood that mental and physical health is worse in the dark months. And most people instinctively know that being cooped up indoors, away from fresh air and light, is unhealthy. So how did we come to override these instincts?

That turns out to be a fairly recent development. A century ago, the association between sun and health was stronger than ever. Then it flipped. How that happened is an intriguing, maddening, and occasionally absurd tale of trends, marketing, mania, and science misconstrued—factors that still plague us today. In the next three chapters, we'll plunge into that story and see how the sun went sideways.

CHAPTER 2

CARE OF THE BODY

The Golden Age of Heliotherapy

On a dazzling December weekend in 1929, a line of traffic left the already-congested downtown of Los Angeles and snaked its way up into the Hollywood Hills. The procession of Model A's and Nash coupes inched past the cafés and movie palaces of Los Feliz and slowly climbed toward number 4616 on Dundee Drive, a narrow cul-de-sac just below the brown expanse of Griffith Park. There, the Angelenos hoped to get a glimpse of the newest and strangest house ever built in the city—a house intended to redefine a new generation's relationship between the self and the elements.

They were coming because they'd been invited by the owner, Dr. Philip Lovell, a naturopath and healthy-living evangelist who authored the popular Care of the Body column in the *Los Angeles Times*. For years, Lovell had been exhorting his readers to embrace the joys of vegetarianism, hydrotherapy, fresh air, exercise, and, above all, sunbathing.

He was preaching to the choir, as a great deal of Los Angeles's explosive growth in the early twentieth century was being driven by those in need of a little more care for their bodies. Diseases such as tuberculosis and influenza had been ravaging Europe and the

United States, and doctors of the time, both mainstream and alternative, agreed that fresh air and sunlight were the most successful treatments.

Los Angeles was a mecca for both. By the 1920s its coastline was dotted with sanatoriums, its downtown filled with pilgrims seeking cures for whatever ailed them. Lovell was the perfect guru for their quest, one of the first in the city's long line of celebrity healers, and in keeping with the modernist moment, he urged his readers to break with their pill-pushing doctors and take charge of their own health, beginning with their living spaces. "A flat, open roof can be of immense value as a sun parlor," he explained in a manifesto in the *Times* in 1924. "The value of sun bathing for anemia, tuberculosis, and practically all of the wasting diseases has been shown repeatedly. . . . There the sick and the suffering may get the benefit of life-giving rays of the sun without the expense of the sanatorium. There man may get in touch with the cosmic forces of nature far better than in closed rooms and confined spaces."

But embracing cosmic forces on your flat, open roof wasn't something the average Angeleno could do in 1924. It wasn't even something Lovell could do. Such houses didn't exist. That didn't dissuade him, however, because a new generation of radical architects had arrived in Los Angeles in the 1920s, full of revolutionary ideas about the role cosmic forces should play in your roof-deck and other parts of your life, and by 1929 Lovell had become their dream client.

"For years I have periodically written articles telling you how to build your home so that you can derive from it the maximum degree of health," he wrote in the newspaper on a Saturday in December. "Always at the end of each article was the thought, 'If I ever build a home myself'—At last the day has arrived." He then did something today's celebrity gurus would not. He invited the entire city to tour the house two weekends in a row. "Consider this an invitation for all Care of the Body readers to visit this newly constructed home built for health." He even drew a map, adding that the house was at the end of a blind street, so please go past the driveway, turn around, and park on the right curb facing downhill.

And they came. By the thousands. What they found was a temple of glass and air, a new vision of the relationship between outside and in, between human and planet, between light and life. The first steel-framed house in the world, rising on impossibly thin supports, it seemed to float above the steep hillside like some houseboat from the future that had traveled back in time and anchored over Los Angeles. There were indeed plenty of flat roofs where one could sunbathe to one's content, a sleeping porch for every bedroom, open-air dining rooms, and soaring walls of windows that looked over the city to the Pacific beyond.

The Lovell Health House, as it came to be known, would become one of the most influential buildings of the twentieth century, and its young architect, Richard Neutra, would come to define California modernism. And though the style is still celebrated for its clean lines, open plans, and spectacular light, people forget that it was driven by health. It was indeed a temple, manifested by a civilization that believed so deeply in the healing power of light that it had chased the sun all the way to the edge of the continent.

It was also an apotheosis. The creation of the Lovell Health House in 1929 marked a high point for heliotherapy, the sun-centric school of medicine that had grown out of the disease epidemics of the 1800s and had swept the medical world. Not many years after the house was created, the medical establishment would begin to turn against the sun with the discovery that its ultraviolet rays could cause skin cancer. With the rise of antibiotics and other pharmaceuticals in the 1940s and '50s, looking to the sun for health would come to seem antiquated. Today, after a multimillion-dollar renovation, the

Lovell Health House stands more stunning and dreamlike than ever. But it is now a monument to a lost world.

TWENTIETH-CENTURY DOCTORS AND architects were far from the first to look to sunlight for its healthful properties. Judging from the frequency of sun depictions in Neolithic rock art, the association is as old as humanity itself. The ancient Egyptians, Greeks, Romans, Arabs, and Indians all used sunlight to treat skin diseases. In the fifth century BCE, for example, Hippocrates had already observed that many diseases followed a seasonal pattern, rising in winter when the sun withdrew, and Pliny the Elder recommended sunlight as "the most powerful of all remedies, and one which is always at a person's own command." The eleventh-century Arabic *Canon of Medicine* suggested that sunlight "invigorates the brain"—something modern medicine is now confirming.

But the scientific elucidation of why the sun might cure disease, and how it could be used in treatment, wouldn't arrive until the nineteenth century, when Europe found itself decimated by two diseases: rickets and tuberculosis. Rickets had been known since classical times, but its incidence exploded in the eighteenth and nineteenth centuries, mystifying doctors. It's an awful disease in which bones don't sufficiently harden, leading to bowed legs, sunken chests, twisted spines, deformed heads. By the peak of the industrial revolution, so many children in England's working-class cities were suffering from it that it came to be known as the English disease.

Why England, the most prosperous country on the planet at the time? This question was first asked by Theobald Palm, a Scottish medical missionary who had been stationed in Asia for many years. Upon his return to England in 1884, he was struck by the prevalence of the disease in England and Scotland, something he'd never seen in Asia. He gathered information from other medical missionaries around the world and confirmed that Northern Europe was the hot spot. In contrast, Spain, Italy, Greece, and Turkey seemed to "enjoy a notable immunity from it."

But Palm also made a more fine-grained geographical distinction. Rates were sky-high in London and Edinburgh—or Auld Reekie as locals called the soot-filled city—but much lower in the Scottish countryside. Palm guessed this was because farm kids were exposed to more sunlight than kids living in tenements and working in factories.

He was right. We now know that sun hitting skin produces vitamin D, a hormone that regulates calcium in the body. D helps ensure that enough of the mineral gets into bone to make it strong. Rickets was appearing in all the cities of the world as they industrialized, but England had the tallest buildings, the sootiest skies, the biggest tenements, and the most kids with rickets. A little sunshine each day would have cured the whole problem, but the doctors of the time had trouble believing that something as immaterial as sunlight could have curative powers, and they didn't pick up on Palm's suggestion.

Not until 1913 would the mystery begin to be solved. By then, Scotland still had the highest rate of rickets in the world. Just a few years earlier, scientists had discovered that cod-liver oil contained a substance that improved growth and health, and they named it vitamin A. Cod-liver oil was also a folk treatment for rickets, so a London doctor named Edward Mellanby wondered if vitamin A might be the key. To find out, he caged beagles in his lab and fed them oatmeal, the standard Scottish diet of the time. Sure enough, the dogs got rickets. (With the benefit of hindsight, we can see that Mellanby got lucky; his dogs got rickets because of the combination of the lousy diet and lack of sunlight.) When he fed them cod-liver oil, it cured their rickets. Success!

Sort of. Elmer McCollum, the University of Wisconsin scientist who had discovered vitamin A, put Mellanby's findings to the test. He bubbled oxygen through cod-liver oil to destroy the vitamin A, then repeated Mellanby's experiment. The cod-liver oil still cured rickets, meaning some substance other than vitamin A was responsible. By that time, vitamins B and C had been discovered, so McCollum named the mystery substance vitamin D.

But technically D should never have been labeled a vitamin. True vitamins are essential substances, required in small amounts, that the body can't produce on its own. Vitamin D was just the opposite: something the body makes with ease from the simplest and most freely available of ingredients. That humans had entered an era when much of the population was unable to complete this basic task should have been a warning sign that we were headed down a dangerous path.

But the era of nutritionism was upon us. Foreshadowing what was to come, doctors opted for a quick fix over a lifestyle intervention. For a while, confusion reigned as to the cause of rickets. One camp argued that it was environmental, being driven by lack of sunlight, while the other argued that it was nutritional, driven by lack of vitamin D, whatever that was. The two theories seemed irreconcilable. But in the 1920s another University of Wisconsin scientist named Harry Steenbock found that he could cure rickets in rats by irradiating either them *or their food* with UV light, and a consensus emerged: vitamin D was produced when certain cholesterol molecules were hit with UV light. It worked in living bodies, and it worked in foods that contained the same fatty cholesterols.

In the following years, some children were successfully treated with UV lamps, but those were rarely accessible to the poorest children who needed them the most. Cod-liver oil ("sunlight in a bottle") was more available, but it was always a tough sell, even when introduced in a new mint flavor that promised to give kids "well-shaped heads." The easiest solution was to increase the D content of common foods by directly irradiating them with UV lamps, or simply adding artificially produced D later on. Fortified foods have been standard fare ever since, and rickets was virtually eliminated—until recently. It has begun to reemerge in cities around the world as a new generation of children are exposed to less and less sunlight—yet another warning, if we care to listen.

HELIOTHERAPY. IT'S A NICE WORD. One feels a little better just saying it. I imagine the first generation of tubercular pilgrims must have felt their spirits lift as they wound their way through the valleys of Switzerland, en route to the Institute for Heliotherapy high in the Alps. There they would settle into their Grand Hotel accommodations and begin treatment in the south-facing solaria with the sliding windows and retractable roofs, covered by sheets except for their feet on day one, then slowly exposing more and more of themselves to the healing rays until no sheets were needed, just goggles and loincloths to protect the dainty bits. By the end of a few months, they were deeply tan and frequently cured.

That was no small feat. From the middle of the eighteenth century to the middle of the twentieth, tuberculosis loomed large over European existence, killing millions. The disease had been simmering on the back burner of civilization for millennia, but the increasing density of populations during the industrial revolution allowed it to catch fire, and it tore through the overcrowded cities of Europe and North America worst of all. By the late nineteenth century it was responsible for one-quarter of all deaths on both sides of the Atlantic.

The bacterium in question could infect the lungs—the classic "consumption" of the Romantic and Victorian eras—or the skin, and this horribly disfiguring flesh-eating form proved surprisingly susceptible to sunlight.

The father of heliotherapy was a sickly doctor named Niels Ryberg Finsen, who was born in the Faroe Islands, a forsaken pile of rocks in the North Atlantic, east of Iceland. Too bleak for trees or crops, the Faroe Islands are a fine place from which to appreciate sunlight, and that certainly was the case for Finsen, who suffered from Niemann-Pick disease, an excruciating malady that results in the slow breakdown of multiple organs. Debilitated by anemia and exhaustion, stuck indoors in a desolate corner of the world, he found himself craving light. "I long for it," he later wrote. "All that I have accomplished in my experiments with light and all that I have learned about its therapeutic value has come because I needed the light so much myself."

As much as possible, he parked himself in the sun and found that it brought some relief. "As an enthusiastic medical man I was of course interested to know *what benefit* the sun really gave," he wrote. "From this time (about 1888) I collected all possible observations about animals seeking the sun, and my conviction that the sun had a useful and important effect on the organism became stronger and stronger."

Finsen became a doctor in Copenhagen, where he began experimentally treating cutaneous tuberculosis with sunlight. The treatments were remarkably successful. We now know that was thanks to the bactericidal qualities of strong ultraviolet light. In 1896 Finsen opened the Finsen Light Institute in Copenhagen, treating tuberculosis sufferers from all over Europe. Originally he used concentrated sunlight on his patients, but he soon invented a carbon-arc lamp that could deliver UV without the other wavelengths and used that extensively.

His techniques became standard, and in 1903 he won the Nobel Prize "in recognition of his contribution to the treatment of diseases, especially lupus vulgaris, with concentrated light radiation, whereby he has opened a new avenue for medical science." Sadly, he couldn't cure his own disease, which had by then confined him to a wheelchair, too weak to attend the ceremony. He sent his regrets, donated the prize money to his institute, and died a year later at age forty-three. But heliotherapy was just getting started.

The same year of Finsen's Nobel Prize, a Swiss doctor named Auguste Rollier founded the Institute for Heliotherapy in Leysin. It was the age of the sanatorium, as chronicled in Thomas Mann's classic novel *The Magic Mountain.* Tuberculosis patients made the pilgrimage to the Alps to cure themselves through a regimen of exposure to fresh cold air, augmented by exercise and a healthful diet. A vision of fin de siècle seriousness with round glasses and slicked hair, Rollier had accompanied a girlfriend with tuberculosis to a sanatorium, where he'd been struck by the extreme health of all the Alpine communities. He believed sunlight was the factor, and he developed his own protocol that augmented the usual treatments with several hours per day of "sun cure." The regimen proved popular, and eventually thousands of patients from around the globe made the journey to his institute, glimmering above the snow at an altitude of forty-five hundred feet.

At first, Rollier focused on external tuberculosis, for which the bactericidal quality of sunlight is highly effective, and he delivered a famously high cure rate. But he and others later extended the treatment to cases of pulmonary tuberculosis as well. The reasoning was less sound, but they did get many things right. They recognized that, in addition to directly killing the tuberculosis bacterium, light seemed to have a tonic effect on the entire body, from muscles to metabolism. They also discerned that morning light seemed to be the most beneficial, and the harsh summer midday sun the least. They observed that those with very pale skin who tended not to tan in response to sun exposure also tended not to benefit much from the treatments. And perhaps most important, Rollier was clear from the beginning that the light itself was just one factor of the cure:

"In tuberculosis, the ultimate issue of a case is to a large extent dependent on the mental condition of the patient, as the struggle with the disease is bound to be a long one, and the patient's courage and endurance are constantly called into play. Some patients possess sufficient determination of character to carry them through the long months of treatment even with dull surroundings; the great majority, however, are largely dependent on their environment to keep up their interest in life and their will to get well."

Fortunately, he had just the place: "The close relationship between sunshine and happiness is so obvious that it hardly requires emphasis. Anyone who has seen the splendour of a typical winter's day in the Alps with its brilliant sunshine and still, cold air will realize what a stimulating effect it has." Indeed, how could one not recover amid such a gestalt? The rare air, the shimmering peaks, the hale and hearty cowherds.

By 1910 Rollier had decided that it was all well and good to treat advanced cases of tuberculosis, but prophylaxis would be even better. "If the environment be sufficiently unfavourable the strongest constitution may fall a victim to tuberculosis," he wrote. "I am also convinced that the converse is true, as my experience in the education, in open air and sunshine, of children coming from families in which tuberculosis has wrought terrible havoc, has shown me that with proper hygienic surroundings the disease can with certainty be avoided."*

* He might have been onto something. Rollier's records don't make for high-quality scientific data today, but in 2025 the dermatologist Richard Weller found that tuberculosis was less than half as common among Brits who got a lot of sun exposure compared to those who got the least. "Even though Finsen and Rollier were not in clinical trials," he says, "they were right."

Rollier launched the School in the Sun for such children. All activities, including school, took place outside year-round. "We reduce clothes to a minimum so as to expose the body freely as possible to the sun," Rollier explained, citing the Greeks' enthusiasm for naked exercise as precedent. Images from his 1923 book, *Heliotherapy*, show the tots sitting at their desks in the snow, wearing only shorts and boots as a similarly clad instructor takes them through their bracing lessons. "The mind of a child who has taken the habit of a disciplined, active outdoor life, will be free from many mean and vile thoughts," Rollier wrote. "In the 'School in the Sun' he is taught the study of Nature's sublime book, he will learn how to love it, and, as a natural consequence, how to love its Creator; and whilst the body gets strengthened, the intelligence opens, the character gets firm, the soul ascends."

BY THEN CHARACTERS WERE firming up all over Europe as individuals, institutions, and governments took a page from Rollier's book. Consumptive French children were sent to the Riviera to convalesce, and the Istituto Elioterapico opened outside San Remo.

Solaria dotted the American landscape, too. The first were founded in the cool, clean air of the Eastern mountains, within reach of the major cities. One of the most prominent was the Adirondack Cottage Sanitarium, founded on the shores of Saranac Lake in 1884 by the New York physician and philanthropist Edward Livingston Trudeau (great-grandfather of *Doonesbury* cartoonist Garry Trudeau). Laid low in New York City by tuberculosis (which had already killed his brother) and told he had just months to live, Trudeau headed north to "bury myself in the Adirondacks," his favorite place.

Month after month, he rested beneath the pine trees beside the lake, breathing the bracing fresh air and waiting to die. To his surprise, he began feeling better and better. Fully recovered, he decided to offer the treatment to others, forgoing the Grand Hotel style of Europe's sanatoriums for individual wooden "cure cottages" outfitted with screen doors, decks, and "Adirondack cure chairs," with angled backs, so his patients could maximize their fresh-air intake in all seasons.

In the Midwest, John Harvey Kellogg incorporated Rollier's heliotherapy methods into the program at his Battle Creek Sanitarium, arguing that "excluding ourselves from the light, we are depriving ourselves of the benefit of the most powerful of all known vital stimulants." The light in Michigan, however, was no match for that in the Swiss mountains, and he soon switched to Finsen-style arc lamps, and then the Electric Light Bath, a kind of mirror-lined cabinet with forty-eight independently controlled light bulbs, which made a big splash at the 1893 Chicago World's Fair.*

In truth, neither Michigan nor the Adirondacks had the climate or light most tuberculosis victims were craving, and those with the means headed West. The high, dry air of the Southwest became a major draw. The Jewish Consumptives' Relief Society lined the porches of its Colorado campus with beds, where its patients took the mountain air in all seasons. Taos was one of the only places D. H. Lawrence ever felt well, and he pined for those desert mountains as he lay dying of tuberculosis in Europe. He made it back only as ashes, interred on his New Mexico ranch. By 1925, the bed count in America's sanatoriums had swelled to 675,000.

But America's largest sanatorium was the state of California itself, whose climate was believed to have potent curative powers

* Kellogg was also one of the first to recognize the value of infrared light, noting that it penetrated much more deeply into the body than visible or ultraviolet light, and that it seemed to improve circulation and the general health of deep tissue.

ever since early pioneers had observed that the indigenous population seemed to possess extraordinary health and longevity. "The climate is so healthy," one doctor noted in 1849, "that illness is rarely found and one sees many persons who have reached an age of over a hundred years." As early as 1870 the state was actually promoting itself as "The Sanatorium of the World." If the first wave of gold rushers had been looking for ore, the second was panning for photons.

Los Angeles, in particular, drew countless health seekers, including Harry Chandler, a tubercular Dartmouth College student who headed to California in the 1880s to clear his lungs and wound up working for the *Los Angeles Times.* Chandler soon married the daughter of the publisher, took over many day-to-day operations of the paper, and became its publisher in 1917.

Always obsessed with physical health and clean living, Chandler came under the sway of the naturopath Philip Lovell, who began treating both him and his wife in Lovell's downtown clinic. Chandler handed Lovell the Care of the Body column in the *Times*, and Lovell didn't shy from the job. His advocacy for exercise, avocados, green smoothies, intermittent fasting, and nature would leave a lasting mark on Los Angeles. "The ideal environment for one with tuberculosis is the country," Lovell told *Times* readers, "especially the dry mountain or desert country where he may bask nude in the sun for hours at a time and where the home environment is such where he may recover his health."

To his credit, Lovell didn't just preach the virtues of clean SoCal living; he and his circle of wealthy progressives put it into practice, starting schools that combined John Dewey's teaching tenets with Rollier's heliotherapy principles. Their goal was to integrate the dreamy California lightscape into every aspect of daily life, to dissolve the walls between indoors and out.

They took their cues from a new generation of architects who were questioning the basic precepts of buildings in light of the waves of disease that seemed to be flourishing in the cloistered cities of the industrial age. The science of germ theory made the trappings of Victorian living look downright suicidal. Tchotchkes harbored dust and germs. Small rooms impeded airflow. It was time to reimagine the twentieth-century dwelling as an extension of the individual's life-support system.

Leading the way was Le Corbusier. Many people know the famous phrase from his 1923 manifesto, *Towards a New Architecture*: "A house is a machine for living in." But they forget the context that precedes it: "The machine we live in is an old coach full of tuberculosis." The clean lines and spaces of modern architecture were a direct response to contagion, a postwar longing for a metaphysical detox.

For their schools and homes, Lovell and his social set commissioned two young architects, Rudolph Schindler and Richard Neutra, who were both students of Frank Lloyd Wright's and were eager to push the boundaries of Le Corbusier's manifesto. Not only could the built environment support an individual's physical health, it could also serve as scaffolding for the spirit. Together, they created a collection of minimalist masterpieces that traded ornamentation for light and embraced the dramatic terrain with a flair that Rollier's stodgy Swiss institute could never have matched.

The most famous of them all was the Lovell Health House, which featured an open-air classroom and spacious interiors lit by vast banks of windows made of vita-glass, a new product designed to let ultraviolet rays pass through. Vita-glass had been created for the Monkey House at the London Zoo in 1925, after it was discovered that the monkeys were coming down with rickets, and it quickly became popular with avant-garde builders—and not just in Los Angeles, where the abundant sunshine made it a bad fit anyway.

Apartment buildings in New York City were retrofitted with the glass, and planned communities in Forest Hills and Scarsdale Downs were built with vita-glass and sunporches to keep families healthy year-round.

That was representative of an international trend. Philip Lovell's helioboosterism may have had a particularly California flair to it, but the naturopath was actually in line with mainstream medical opinion of the time. Everywhere doctors looked, industrialized landscapes seemed to be causing disease, and the sun seemed to be rooting it out. Had we turned our back on nature too soon?

IN WORLD WAR I, surgeons working under brutal conditions found sunlight to be an invaluable tool. In a 1916 article on advances in war surgery, *Scientific American* observed that "even the most serious and extensive wounds will heal much more rapidly" in the presence of sunlight. That was part of a growing general understanding that disease lurked in the dark and sooty, the trenches and the mines, while health held in the bright and airy. Sunlight began to morph in the eyes of the medical profession from a tool for specific conditions to a general bulwark of good health. Schoolhouses were redesigned with high banks of windows to improve students' learning and health and to prevent myopia.* Hospitals wheeled their patients onto the roof to absorb the healing rays of the sun, even in winter, and wards were redesigned to ensure a steady supply of light and air.

In England, heliotherapy's biggest booster was the physician and health activist Caleb Saleeby. "The treatment of disease by sunlight

* And it works. A number of studies have shown that kids retain information better and score higher on tests in classrooms illuminated by natural sunlight. Rates of myopia are also vastly lower in kids exposed to more natural light.

is the newest of old things," he wrote in *Nature* in 1922, introducing Auguste Rollier's work to the anglophone world. "We shall do well not to associate the sun cure exclusively with tuberculosis; nor solely with the proved antiseptic power of sunlight. The recent American work, both at Columbia University and Johns Hopkins Hospital, has shown that sunlight has potent influences upon nutrition and metabolism."

For Saleeby, the big issue was the smoke clogging the world's cities, which sickened people both directly and by blocking the sun. He founded the Sunlight League in 1924 to spearhead a movement for cleaner skies. "The air is abominably befouled, and the light of life is blackened by the darkness," he warned attendees to the First International Conference on Light (held in the clean air of Rollier's Institute for Heliotherapy in 1928). "For lack of the primal necessity, sunlight, we die in vast numbers far exceeding those due to any other cause."

Throughout the 1920s, sunlight was championed by the medical establishment as the best way to improve public health. "Light is life, and lack of it causes ill health," the celebrated British pediatrician Helen Mackay told *The New York Times* in 1924. "The therapeutic values of sunlight can scarcely be overestimated." In 1925, *The Journal of the American Medical Association* confirmed that "the antirachitic effects of exposure to sunlight discovered during the last few years indicate the therapeutic and prophylactic efficacy of sunlit air." The US government distributed a pamphlet titled *Sunlight for Babies*, which urged sunbathing for both expectant moms and for kids.

In 1926, *The Lancet* called sunbathing "one of Nature's greatest aids to maintaining and acquiring proper health." Cornell was one of many institutions to establish a solarium for its students. Doctors readily agreed that the enemy of health was indoor life. "No deficiencies that children develop are of greater significance than those caused by the lack of sunlight," the president of the Chicago Board of Health wrote in *Ladies' Home Journal.* "The sun is the constant mortal enemy of germs. . . . Let us get out into the sunshine with the children as often as possible."

No country took that exhortation more to heart than Germany, where *Nacktkultur* (nudism) became an unlikely national passion. Whole families flocked to their local *Freilichtpark* (free-light park) for group calisthenics under the sun's salutary rays. That paled in comparison to Moscow, where on sunny days half the city could be found in their birthday suits on the banks of the Moskva River, soaking up the rare photons.

The movement found a pliant ally in France, but it was always a tougher sell in chilly Britain, despite the efforts of reformers such as Caleb Saleeby, whose tireless advocacy could never convince more than a few Brits to strip off in the sun. The nadir came in 1930, when a community of nudists who had commandeered London's Welsh Harp Reservoir were assaulted by a mob shouting, "Hottentots would behave with more decency!"

The incident triggered a brief national soul-searching. When the London *Times* published a letter calling for the establishment of public places where London's citizens could engage in "active air-bathing," *The Lancet* weighed in with its support: "On first consideration, the idea of a community of people deliberately practising nudity, especially with municipal encouragement, strikes the average person as somewhat ridiculous. . . . But the discovery that the rays of the sun on the skin exert a beneficent effect on health has done something to undermine these prejudices. The thought of tuberculous children, now brown-skinned and active, tumbling about in Alpine snows is no longer disturbing, nor even the prospect of adults doing these things, as long as they are done for the sake of health and as long as the sexes are segregated."

For a surreal moment, it looked as though nudism might sweep the world. The American writer and social critic Stuart Chase knew he was preaching to the choir when he published "Confessions of a Sun-Worshiper" in *The Nation* in 1929. "Some people collect postage stamps, others, old masters," he began. "I collect ultraviolet rays." Chase's lifelong habit began years earlier at the old L Street Bathhouse in South Boston, where he discovered and quickly joined hundreds of naked men from the city's ranks taking the rays year-round. "In a way it is like a drug," he noted perceptively. "The after-effects are a sense of well-being, of calmed nerves, of inner vitality."

The year 1929 was also when the Lovell Health House made its mark. It truly looked as though the whole country might be swayed to sun. "America, we shall undress and bronze you yet!" Chase wrote. "I have been associated with many reform movements in my life, and it is with considerable astonishment that I find one actually gaining ground. . . . If the republic wants to go native and can hold to it with any fidelity, it will probably do more than any other conceivable action to balance the inhibitions and pathological cripplings induced by the machine age and the monstrous cities in which we live."

By 1930, the association between light and health had infused modern culture to an intensity that has never been matched. Bathing suits shrank, hotels tricked out their roofs into sunbathing destinations, and even the breakfast cereal Wheatena felt it necessary to point out on its box, "The same *natural* rays of the sun that brown your skin—that pour health into your body—also give Wheatena its color and its wholesomeness." America had turned as golden as a field of August grain.

IT'S EASY TO MOCK the excesses and naivety of the heliotherapy era, to laugh at the thought of instructors and their students doing math in their underwear in the Swiss snow, but woven into the silliness were some solid instincts. Bodies obviously benefited from light and air and a range of temperatures and exposures, from mixing it up with the physical world in ways that kept them competent. That sentiment had been growing since the turn of the century, with the birth of a new generation of women eager to throw off the corsets, creams, powders, and parasols of the Victorian era and let a little light in, as *Harper's Bazaar* observed in 1900:

> The summer girl of 1900 is ready to take a spin in an automobile; or to speed forth on her bicycle; or to hold her own with a racquet in her hand at the tennis nets; or with her sticks to speed her ball over the short or long course of the golf-links; or to take her ocean bath, and with sturdy strokes to swim and disport herself like a mermaid in her abbreviated bathing costume; or to row and sail and yacht from early morn until late at night, letting the sun leave what impress upon her it can or may.

When we look back, that sounds like a thoroughly modern approach, and one can envision an alternative history where it all went sensibly. Human health flourished out of doors and the sun played an important role, but it was also clear that a little went a long way. Even as doctors were getting their first inklings of the sun's carcinogenic dark side, there was still a chance to steer a course toward moderation.

Unfortunately, that's not how trends go. If a little sun is good, then a lot must be even better. Even burning came to be associated with benefits. The beaches turned into "public rotisseries," complained *The New York Times*, which dryly observed that people "took to sun baths with an avidity they had never shown for spinach, sleep, or orthopedic shoes." *Collier's* dubbed it "ultraviolet insanity." Coco

Chanel, who often gets credit for the trend, was actually late to the party, though she certainly helped goose it with her daring beachwear and bronzed Saint-Tropez skin. "A golden tan is the index of chic!" she declared in *Vogue* in 1929, making it so.

The market took charge from there, inventing products to sell people so they wouldn't have to rely on something as inconvenient as nature. Who needs the beach when modern technology allows you to get UV year-round in the comfort of your own home? Stores filled with UV machines that looked like ray guns from a *Flash Gordon* comic, but in this case shot "rays of health . . . flooding your whole body with energy and well-being."

The wildly popular lamps had the blessing of the American Medical Association, which was mostly concerned that they shouldn't emit any wavelengths of radiation that wouldn't naturally be found in sunlight, though some of them did and were probably quite dangerous. General Electric promised that its lamp could deliver a "mild sunburn" in just twenty minutes, and the company urged parents to "give children ultra-violet radiation in their playrooms." Its chief engineer, Matthew Luckiesh, predicted, "In the future . . . we shall sleep in beds shaped as covered wagons, and instead of pajamas we shall have ultra-violet light rays pouring down on us as we sleep, producing the effect of reposing on a sunlit meadow in a tropical land. The benefit to our health should be incalculable." Schools gave up on natural lighting and instead dosed their kids with ultraviolet lamps every day. The US House of Representatives even added a sunlamp to its shower room.

Once again, we'd missed the point. When some complex natural ingredient or behavior is reduced to a pill or product, the benefits tend to get lost, though we rarely learn the lesson. Blasting yourself with a UV gun all evening was a long way from incurring some incidental sun exposure while swimming or gardening, and medical authorities grew increasingly nervous about a possible connection between UV and skin cancer.

At first, they were too uncertain to make any bold public pronouncements, instead offering measured suggestions to not overdo it. But as people continued to fry themselves with abandon, even as scientists were learning more about the dangers, a sense of urgency took hold, which laid the ground for a familiar dynamic. For their own good, people had to be broken of their sun addiction. If nuanced explanations didn't work, then they'd have to be scared straight.

CHAPTER 3

SLIP, SLOP, SLAP

The War on Skin Cancer and the Rise of Sunscreen

Considering the sun's ability to blister skin, you'd think its link to skin cancer would have been more obvious, but it took medical men a surprisingly long time to figure it out. The German dermatologist Paul Gerson Unna had an inkling in 1894, noting sailors' high incidence of carcinomas in his textbook and pointing to the extreme sun exposure of the occupation. A few others picked up on the connection, including the Chicago dermatologist James Nevins Hyde, who wondered if ultraviolet light might be an incriminating factor in his 1906 article "On the Influence of Light in Production of Cancer of the Skin."

Yet these guys were outliers, and even they were not wholly convinced. Most people, after all, didn't develop skin cancer, even those with chronic sun exposure. The connection was further obscured by the long gestation period, which can be decades. At most, the experts assumed sunlight might be one factor of many.*

* And they were right. We now know that everything from diet to obesity affects your risk of skin cancer. (Diet may have been a factor in the high rate among sailors, who get few fruits and vegetables.)

As late as 1929, the notion that sunlight can trigger skin cancer had made few inroads with the medical mainstream. When questioned about the connection that year, even *The Journal of the American Medical Association* responded by stating, "There is no evidence available that exposure to the sun predisposes to epithelioma of the skin."

But there was. Just the year before, England's George Marshall Findlay had zapped albino mice with high doses of UV light from a mercury-vapor lamp, which produces light by running electricity through a glass tube of mercury. When energized, the mercury atoms produce a lot of UV, as well as visible light. In normal settings, the UV is filtered out by glass, but mercury-vapor lamps can also be used in the lab to test the effects of UV light. Findlay's mice sprouted tumors all over their skin. "These experiments, therefore, are in favour of the hypothesis based on clinical evidence that in man exposure to ultra-violet light plays an important part in the aetiology of cancer of the skin," he announced in *The Lancet.*

Clearly this was news even to most doctors, but the one report doesn't seem to have been enough to inspire any action on their part. The American Medical Association put most of its focus in the 1930s on preventing the hype over sunlamps from getting out of hand. In addition to issuing its approval only for lamps that produced a spectrum and intensity similar to that of natural sunlight, it refused to endorse any exorbitant health claims by the manufacturers, recommending that phototherapy be used only for rickets and the many skin conditions dermatologists were already using it for, including psoriasis, eczema, acne, and tuberculosis.

But the more doctors began to look for connections between excessive sun exposure and skin cancer, the more they found them. "Every physician knows that farmers and seamen are especially prone to develop skin cancer," James Ewing, director of Memorial Hospital in New York, told *The New York Times* in 1933. The signs were just as obvious in hardcore sunbathers with fair skin, especially those who lived in very sunny places such as Australia. Doctors became more cautious in their suntan messaging and conveyed this caution to the popular press. In the 1940s, magazines including *Life*, *Ladies' Home Journal*, and the *Saturday Evening Post* ran articles warning that excessive sunbathing likely raised the risk of skin cancer.

But even in the medical profession, there was little agreement on the magnitude of that risk. Addressing a forum of doctors at the American Cancer Society in 1941, Harold Blum, who had cofounded the National Cancer Institute three years earlier, was still trying to make the case to his peers that UV could induce skin cancer. He concluded, "The evidence is such as to constitute a warning against excessive exposure to sunlight. At the same time it does not justify fear of moderate exposure."

That was the party line through the mid-twentieth century. If you have a job that keeps you in the sun, or a predilection for hitting the beach every day, you need to worry about skin cancer. But don't sweat the small exposures.

That position would, however, eventually be marginalized by other trends sweeping the culture.

THE 1940S ALSO SAW the rise of a new product that would become a powerful player in the war on skin cancer. The first effective sunscreen was developed in 1942 after the air force came to General Electric's Lighting Research Laboratory with a problem. World War II was raging in the South Pacific, and airmen who wound up on life rafts were getting dangerously fried. The air force wanted a sun-protection product that could be stored on every life raft. In addition to being protective, the product had to withstand the freezing temperatures at high altitude, had to stay on through repeated dunkings in the sea, and had to be as nonbulky as possible.

The director of GE's Lighting Research Laboratory was Matthew Luckiesh—the same scientist who had developed GE's UV lamps a decade earlier. He was the top guy on UV, and he took the job with enthusiasm. Luckiesh tested dozens of substances, from lanolin and zinc oxide to petroleum jellies of different thicknesses. The most promising candidates were tested on volunteers exposed to the UV equivalent of twenty hours of summer sun, then further tested in repeated washing experiments.

The winner was a dark red veterinary petroleum jelly made by Standard Oil Company. It blocked UV rays, was hard to wash off even with soap, and produced no irritation when used for days in succession. It won no awards for smell or style, but if you'd just been shot down in the South Pacific, that was fine. Red vet pet, as it came to be known, was standard issue on life-raft kits for the rest of the war.

After that, an airman named Benjamin Green who had become a Miami pharmacist found himself fixated on red vet pet. Green was bald and was sick of burning his head. He wondered if he could create some version of red vet pet that would do the job without the obvious social drawbacks. After much tinkering, he hit upon a version using cocoa butter and coconut oil. Coppertone Suntan Cream was born.

As implied by the name, those early generations of sunscreens were focused on preventing burning, not skin cancer. The ultraviolet light that reaches Earth's surface comes in a range of wavelengths from 280 to 400 nanometers. By the 1930s, scientists had recognized that some of those wavelengths had very different physical properties, so in 1932, at the Second International Congress on Light in

Copenhagen (just down the road from the Finsen Light Institute, which hosted a field trip), three hundred light researchers decided unanimously to divide the ultraviolet spectrum into three categories: UVC (shorter than 280 nm, which doesn't reach the surface of Earth); UVB (280–315 nm); and UVA (315–400 nm). The categories have stuck ever since.

Sunburns are caused primarily by radiation in the UVB range, while tanning is triggered by UVB, UVA, and higher-energy visible light. Conveniently, the chemicals available to early sunscreen manufacturers blocked UVB while allowing longer wavelengths to pass through. Starting in the postwar era, sunscreen became popular primarily as a tanning aid that could prevent sunburn, allowing you to spend all the more time bronzing.

By the 1970s, however, doctors were growing more alarmed by the skyrocketing rate of skin cancer. Diagnoses went from 300,000 per year in the 1970s to 400,000 per year in 1980 and 500,000 in 1985.

This was actually less of an emergency than it seemed. One of the funny things about cancer is that its incidence rises as health care improves. It occurs mostly in the elderly, so when life expectancy improves and fewer people die from such things as infectious diseases and heart disease, that leaves a lot more old people to get cancer. Throughout the twentieth century, longer lifespans and massive population growth resulted in lots of cases of skin cancer, but this was actually a sign of progress.

In addition, increased surveillance played a role in the rising numbers. In the past fifty years, the number of dermatologists in the United States has risen by 500 percent, much faster than the overall population, and the percentage of people getting screened for skin cancer has steadily risen as well. So the numbers were driven by a huge new generation of elderly people who were getting basal cell carcinomas (which comprise about 80 percent of skin cancer cases and are the least worrisome) and were getting them checked.

Still, the changes in demographics and procedures don't negate that there were lots of cases of skin cancer to be dealt with, and authorities in many countries began stepping up their efforts to get the public to take sun protection seriously. In 1980 Australia, where the problem was most acute, rolled out its Slip, Slop, Slap campaign (slip on a shirt, slop on some sunscreen, and slap on a hat), complete with a PSA cartoon and *Schoolhouse Rock!*–style song featuring Sid the Seagull. The UK and United States were not far behind. Soon both the authorities and the media were working hard to make tanning uncool.

This might have seemed like the death knell for an industry that had hitched its wagon to tanning culture, but it turned out to be a godsend. Whatever sunscreen might have been worth as a tanning aid, it was worth far more as a shield against death. It all seemed quite serendipitous: The same UVB rays that caused sunburn were the ones that directly damaged DNA and caused skin cancer, so people could simultaneously prevent sunburns and skin cancer while tanning to their heart's content. (Or so they thought. . . . We'll come back to this soon.)

By then the industry had been in a weird dance with the FDA for years. If it claimed its products prevented sunburns, then they were a drug and had to stand up to regulatory scrutiny. But if it just positioned them as tanning aids, then they were a cosmetic and could use whatever ingredients they wanted. The companies had always tried to stay on the cosmetic side of the line, but the new focus on skin cancer in the 1970s made it clear that the big bucks were on the drug side. In the late 1970s, after years of lobbying, the companies finally convinced the FDA to allow them to claim that use of sunscreen *might* help reduce the risk of skin cancer.

But what really kicked the business into the stratosphere was the introduction around 1980 of the SPF system, again after years of coordination between the industry and its minders. Pre-SPF, the US sun-care market was about a $200 million industry. Cute, but a creamy drop in the $15 billion bucket it is today. SPF transformed a product that had all the gravitas of surf wax into an anchor of wellness with endless options in strength, formulation, and price. The companies didn't even disguise their intentions. One vice president of marketing told *The New York Times* in 1986 that the goal was to make sunscreen something "you put on every morning when you put on makeup or brush your teeth."

When it released its initial guidelines, the FDA assumed there would never be a need for an SPF higher than 15. The SPF number indicates how many times longer you can stay out in the sun before burning. So if the day is such that you could spend 30 minutes sunning before burning, with SPF 15 you could spend 450 minutes. The FDA's perspective was that no sane person would ever want to waste more than seven hours prostrate in the sun. "There is not enough ultraviolet light on Earth to warrant an SPF over fifteen," said the director of its cosmetics division in 1987. More than that, he said, would be "gilding the lily."

He wasn't thinking like a marketer, of course. The gilded lilies flew off drugstore shelves. If 30 contained twice the cancer protection of 15, that seemed like a bargain for a few bucks more. Heck, why not go for SPF 100, the Cadillac?

The absurdity becomes clear once you consider the numbers. The reason it takes thirty times longer to burn wearing SPF 30 is because it blocks all but one-thirtieth of the UVB rays, or 97 percent. Beyond that, it's diminishing returns. SPF 50 blocks 98 percent. SPF 100 blocks 99 percent. And getting those final, tiny bumps in coverage means using more of the chemicals that block the UV, which are absorbed by the body in surprisingly large quantities and have raised multiple flags at the FDA.* (I'll return to that issue in the final chapter, "Right with Light.")

Initially, the FDA tried to prevent companies from labeling their products higher than SPF 15, but the companies kept pushing hard and eventually the FDA gave in to the inevitable, though it continued to gripe about it, calling SPF 50 "absurd" in 1989.† But regardless of the fine print, as public health campaigns go, the push for 24/7 sunscreen in medical offices, billboards, and media appearances has been a home run. Sun awareness infused the general consciousness, and sunscreen sales went through the roof.

But weirdly, so did rates of skin cancer. And sunscreen manufacturers were having a great deal of trouble explaining why.

HERE IN THE 2020S, decades of consistent messaging has made us accustomed to thinking of sunscreen as an indispensable tool for staying alive. It's natural to think that its benefits have long been obvious to all parties. So it can be surprising to look back and discover how skeptical the experts have been. In 1998, cases of skin cancer in the United States hit 1 million per year and showed no sign of slowing down, and when *Consumer Reports* did its periodic roundup of sunscreens, it wasn't enthusiastic. "The details of the

* They also don't always work very well. In 2025, the Australian consumer-advocate watchdog CHOICE tested twenty leading SPF 50 sunscreens to see if they lived up to their claims. Only *four* did. Most delivered functional SPFs in the 20s and 30s, and one (which was also the most expensive, at about $35 for a small tube) was only an SPF 4. Since then, dozens of sunscreens have been pulled from the market in what the BBC terms the "Australia sunscreen scandal," and more are expected to follow. So take those SPF 50s with a grain of sea salt.

† An argument can be made that people don't apply sunblock as thickly as they are supposed to, so an SPF 50 is really more like an SPF 25 as used in real life, but even that 25x indicates a fear of the sun out of all proportion to the threat.

sunlight-cancer connection are unclear," it reported, adding, "It turns out to be hard to show that people who wear a lot of sunscreen have a lower cancer rate than other people with similar sun exposure."

Hard to show indeed. In fact, most studies showed just the opposite: Sunscreen use was associated with *higher* rates of skin cancer, not lower. This was profoundly inconvenient for the skin-care business, which tried to explain it away in various ways. Maybe people who used more sunscreen were also spending more time in the sun? Or maybe people with very fair skin, who are also more susceptible to skin cancer, were the ones using all the sunscreen?

Fair points, but they can't account for all of the mystery. When the *European Journal of Dermatology* published a meta-analysis of the twenty-nine studies of sunscreen use and skin cancer that had been done since the 1970s, all of them showed a higher risk of skin cancer among the sunscreen users, even the ones that adjusted for skin tone, time in the sun, and other confounding factors.

But the trends were telling. The studies from the 1970s showed that sunscreen users were more than twice as likely to get skin cancer, the ones from the 1980s showed that they were 26–85 percent more likely, and those from the 1990s onward showed almost no difference. This points to problems with those earlier sunscreens that were fixed in later formulations.

And that, as it turns out, is something scientists have been worried about all along, especially regarding melanoma, where it matters the most. Those concerns mostly stayed below the radar until 1993, when the journalist Michael Castleman published an exposé in *Mother Jones* magazine. "The public health authorities who urge routine, liberal use of sunscreen (especially on children) fail to mention that sunscreens have never been shown to prevent melanoma," he wrote. "The medical research community knows this. The Food and Drug Administration knows it. And sunscreen makers know it. Yet, as a result of scientific myopia, bureaucratic inertia, and the almighty bottom line, they've essentially told us to use sunscreen and not to worry." Castleman even quoted a manager from Schering-Plough, owner of Coppertone at the time, admitting, "We don't know if sunscreens prevent it."

Clearly Castleman and *MoJo* were gunning for a fight, but they'd indeed found a weak spot. The ability of sunscreens to prevent melanoma has always been debated. A 1990s meta-analysis of the fifteen high-quality studies that had been completed at that time found that three showed lower rates of melanoma among sunscreen users, four showed no difference, and eight showed higher rates of melanoma. In 2006, the US EPA admitted, "There is no evidence that sunscreens protect you from malignant melanoma." In 2009, the UK sunscreen expert Brian Diffey said the same, writing, "Sunscreens have been shown to be effective in reducing the incidence of squamous cell cancer and with promising benefits for basal cell cancer. However, the evidence that they are effective in melanoma remains lacking."

That's still an issue today. The oncologist Sara Gandini acknowledged in 2020, "There is no high-quality experimental evidence on the efficacy of sunscreen to prevent melanoma," though she suggested we use it anyway in tandem with other more proven measures such as protective clothing. A 2023 study by dermatologists from Harvard Medical School graded the evidence that sunscreen use reduces the risk of squamous cell carcinoma "strong"; that it reduces basal cell carcinoma "not as strong"; but as for melanoma, "the evidence is very unclear . . . as there are even several studies that find a positive association between sunscreen use and melanoma development."

Why would melanoma be any different from the other skin cancers? The notion isn't intuitive to the general public, which tends to lump together all the skin cancers. But they are very different diseases. And in all the ways that the carcinomas are banal and predictable, melanoma is just the opposite.

CHAPTER 4

THE MYSTERIES OF MELANOMA

What Do We Really Know About the Most Dangerous Skin Cancer?

Part of the reason people tend to conflate carcinomas and melanomas in their minds is because of a little switcheroo that the messengers like to pull. When they want to impress you with how likely you are to get skin cancer, they use that broad term. ("One in five Americans will develop skin cancer in their lifetime.") But when they want to drive home how deadly it can be, they talk about melanoma. ("Patients with metastatic melanoma have a poor prognosis, with a median survival duration of only six to nine months.")

That sounds really scary! If I didn't read the fine print, I might think that bad skin cancers are both common and deadly, which is not the case. The common ones aren't very deadly, and the deadly ones aren't very common. Melanomas represent just one out of every fifty cases of skin cancer, and half of them aren't invasive.

Not only are the various skin cancers not interchangeable, they are wildly different diseases. Squamous cell carcinoma occurs in the outermost layer of skin cells, which are constantly getting sloughed off, while basal cell carcinoma occurs in the cells below them that divide to produce new skin cells. Melanoma is a cancer of the melanocytes, the melanin-producing cells, which sit with the basal cells along the lowest layer of the epidermis.

The behaviors of the various skin cancers are as different as their origins. Squamous cell carcinoma, for example, acts exactly as you would expect a sun-induced skin cancer to act. It shows up on the most exposed parts of the body, such as the face and neck. It's most common in those who have had the most cumulative sun exposure. Its tumors bear the classic genetic fingerprints of UV-induced mutations. And sunscreens significantly reduce its incidence.

Melanoma is a different beast. It tends to appear on parts of the body that don't get much sun exposure, while being less common on chronically sun-exposed parts such as the head and hands. It often shows no sign of typical UV-induced mutations. It's more common in office workers than outdoor workers. And it's rare in people who don't have pale skin.

And while melanoma isn't closely associated with lifetime sun exposure, it is with a history of sunburns, especially during childhood. Migration studies show that where you spent your childhood has a much larger impact on your lifetime risk of melanoma than where you spend your adulthood. Harvard's famous Nurses' Health Study found that women who had experienced five or more blistering sunburns before the age of twenty had more than twice the rate of melanoma compared to those with no burns, but found no significant association between burns after age thirty and melanoma. A later study of nearly one hundred thousand French women came up with similar results: no association between burns after age twenty-five and melanoma.

The biggest risk factors are things you can't do anything about: red hair, pale skin, freckles, and lots of moles, which are clusters of melanocytes that occasionally turn cancerous. Only 4 percent of melanomas occur in people with fewer than ten moles.* Harvard dermatologist David Fisher has shown that the gene variant responsible for red hair and freckles also triggers melanoma completely independently of ultraviolet light exposure. So, while sun exposure might exacerbate the problem in fair-skinned people, the root of the trouble lies in the genes.

That idea was confirmed in a 2025 study by Australia's skin-cancer experts that assessed all the factors that predict a future melanoma diagnosis. The biggies were hair color, skin color, freckles, and moles. Height and sex were also good predictors (it's more common in men, especially taller men), and surprisingly, smoking reduced the risk of melanoma by 20 percent. No one knows why.

One of the weirdest things about melanoma is that the survival rate is worse among those who receive little sun exposure. That was pointed out in a 2005 *Journal of the National Cancer Institute* study that famously kicked the hornet's nest and led to several follow-up studies, which found that sun seekers tend to get milder melanomas than sun avoiders, are less likely to suffer a recurrence, and are less likely to die from their melanomas. It may be a disease that is sparked by light but festers in the dark.†

* Diet plays a role, too. A French study of sixty-seven thousand women found that strong adherence to a Mediterranean diet reduces the risk of melanoma by 28 percent. Vegetable intake was the aspect of the Mediterranean diet most strongly associated with lower melanoma rates, which fits with the prevailing theory that melanoma is driven in part by damage caused by free radicals, which can be prevented by maintaining a high level of antioxidants in the skin. Lycopene, the red-pigmented antioxidant found in tomatoes and other foods, is not only associated with lower rates of melanoma but also with a 40 percent reduction in sunburn itself.

† I learned one of the most mind-blowing explanations for why this might be in a 2024 presentation by the UK skin-cancer expert Amaya Viros, who found that melanomas use collagen—the fiber-like protein that provides structure in the skin and internal organs—as a kind of highway to enter the body. So melanomas that arise on supple skin may have an easier time invading the body, while ironically those that arise on sun-damaged skin that has had its collagen destroyed are pretty much stuck in place.

As with the other skin cancers, part of the rise in melanoma rates in the second half of the twentieth century was due to increased awareness and longer lifespans. But melanoma tends to occur at younger ages than the other skin cancers, so that couldn't explain everything doctors were seeing. Age-adjusted rates among White Americans, for example, rose from 10 per 100,000 in 1975 to 30 in 1995. Melanoma was clearly very different from the other skin cancers, and all the slipping, slopping, and slapping in the world wasn't slowing it down. By then, a growing number of researchers had a theory why—a theory that was big trouble for the sunscreen industry.

AS I MENTIONED, the original chemical filters used by sunscreens were designed to block the UVB rays that cause sunburn. UVB was also believed to be the primary radiation responsible for skin cancer because it had the right wavelengths to inject its energy directly into DNA and distort the molecules, causing mutations. At the time, the industry had no chemical filters capable of blocking the longer UVA wavelengths, but that didn't seem to be a problem. UVA has much lower energy than UVB, and it has little direct impact on DNA.

But in the 1980s scientists realized that UVA could also cause cancer. Instead of frazzling DNA, it produces what are known as reactive oxygen species, which are a type of free radical, a molecule that has been energized into a highly charged state. To get back to a neutral charge, it needs to bond with electrons, and it will snatch those electrons from neighboring molecules, unbalancing them and causing a chain reaction of molecular maiming. The result is cellular damage that can lead to cancer.

UVA, it turns out, is good at charging up free radicals. By the late 1980s, this was common knowledge among the experts. "Both UVA and UVB are carcinogenic," one Harvard photobiologist told attendees at the first World Congress on Cancers of the Skin in New York in 1983. There was even good reason to suspect that UVA might be the bigger factor for melanoma because melanocytes are found deep in the epidermis, and UVB rarely penetrates that far. For every photon of UVB that reaches a melanocyte, seventy to one hundred photons of UVA do.

That was the argument of the brothers Frank and Cedric Garland, two University of California epidemiologists who were famous for their work connecting vitamin D deficiency to cancer. At a 1990 conference, they argued that UVB-blocking sunscreens might be unintentionally promoting melanoma, then made the cardinal sin of speaking openly about it to a *New York Times* reporter. "When you block out ultraviolet B light, you stop the skin from burning," Frank Garland told the *Times*. "Which means you can stay out in the sun for many hours or many days. So what happens is you get an unnaturally high dose of ultraviolet A. Sunscreens give you a false sense of security."

The ensuing article bore a title straight out of Coppertone's worst nightmare: "Theory Hints Sunscreens Raise Melanoma Risks." Cue the panic in the industry, which countered with some strong denials that UVA was a concern and accused the Garlands of grandstanding.

But the Garlands pushed their case in a 1992 letter in the *American Journal of Public Health*, pointing out that there was no evidence that PABA, the sunscreen filter du jour, was capable of preventing melanoma, and in fact there was some evidence to the contrary: "Worldwide, the countries where chemical sunscreens have been recommended and adopted have experienced the greatest rise in cutaneous malignant melanoma." Part of the problem, they suggested, was that one of vitamin D's main functions in the skin was to prevent cancer cells from proliferating, so by blocking UVB you were eliminating one of the body's main defenses. The following year, their research was picked up by Michael Castleman for his *Mother Jones* article, though the Garlands had learned their lesson and declined to be interviewed.

"Broad spectrum" sunscreens that blocked both UVB and UVA didn't start to appear until the 1990s and didn't become commonplace until the 2000s. That leaves the 1980s as an awkward window in which many people in the industry must have known about the risks of UVA but continued to provide consumers with products that did nothing to block it.* As the authors of the meta-analysis in the *European Journal of Dermatology* put it, those beachgoers dutifully applying their UVB-blocking sunscreens, believing that no burn meant no risk, "may stay longer in the sun because of reduced risk of sunburn, and thus increase their exposure to UVA radiation which may lead to an increased risk of carcinogenesis."

I was one of them. As a teenager in Florida in the 1980s, I spent loads of time at the beach smelling like a coconut.† I must have absorbed staggering quantities of UVA.

By 1993, when *Mother Jones* did its exposé, things were looking downright dicey for the sun-care corporations. The article quoted a vice president for Tanning Research Laboratories, owners of Hawaiian Tropic, admitting, "When you protect only for UVB, it would seem to run a risk for potential skin cancer. UVA is a more damaging ray. We may be hurting ourselves by protecting ourselves too well on the UVB side."

And it kept getting worse. In 1994, a team led by Margaret Kripke, a pioneering cancer researcher at MD Anderson Cancer

* To this day, SPF applies only to UVB, and UVA blockage can be spotty.

† We favored Hawaiian Tropic, the most coconutty of them all. Little did we know the first batch had been mixed up in a trash can right there at Daytona Beach in 1967 by a lifeguard who decided to take the coconut to eleven.

Center in Texas, applied sunscreen to the ears and tails of mice, exposed them to UV, and found that it prevented sunburn but had no ability to prevent melanoma. In her paper in the *Journal of the National Cancer Institute,* Kripke wrote, "Sunscreen protection against UV radiation-induced inflammation may encourage prolonged exposure to UV radiation and thus may actually increase the risk of melanoma development."

Kripke's study drew coverage in *The New York Times*, under the title "Mouse Study Raises Doubts on Sunscreens." The piece acknowledged, "It has proved extraordinarily difficult to unravel the relationship between sun exposure and melanoma and that in this respect melanoma research stands in stark contrast to the well-established findings tying sun exposure to less deadly skin cancers, the basal and squamous cell carcinomas." (That prompted a triumphant letter from the editor in chief of *Mother Jones*, saying essentially, "We told you so.")

Sensing a potentially disastrous situation, the powers that be closed ranks, insisting that there was no evidence that UVA could cause melanoma or that sunscreen might exacerbate the situation. But there were signs of extreme nervousness in the industry. Donald Davis, editor of *Drug & Cosmetic Industry* magazine, opined to his colleagues in 1994 that Kripke's experiment "should cause red flags to be run up in the offices of sunscreen marketers." In light of the new evidence, Davis wrote, "The honeymoon for sunscreens is about to experience some serious scientific introspection."

And yet that introspection never happened. In the nick of time, the industry came up with avobenzone, a chemical filter that could block UVA, at which point it pivoted from downplaying the dangers of UVA to promoting the importance of broad-spectrum sunscreens that could block it. That stance continues today. Here, for example, is the industry-funded Skin Cancer Foundation website in 2025: "While we used to think UVA light mostly just caused skin aging, we now know that its longer wavelength penetrates the skin more deeply and is strongly linked to melanoma."

That's cold comfort to the sunscreen users of the 1970s and '80s. And it possibly explains part of the terrible pattern in that era, as rates of melanoma continued to skyrocket despite a gusher of sunscreens of higher and higher SPF.

In 2010, the industry finally got the study it had been waiting for. In 1992, researchers in Nambour, Australia, assigned about 800 participants to a daily regimen of broad-spectrum SPF 16 sunscreen, while another 800 or so were told to continue with their usual sunscreen routine. All participants were tracked as they went about their daily lives for the next four and a half years, which in supersunny Queensland probably meant a fair amount of exposure. Their health was followed for an additional ten years. Only 11 of the daily-sunscreen users were diagnosed with melanomas, versus 22 of the discretionary-sunscreen users. A 50 percent drop is a big deal.*

Being able to point to that study was a godsend for the sunscreen industry. The Australian Academy of Science literally refers to it as "the study that saved our skin." It remains an outlier, however, a single study struggling to support the weight of an entire medical-industrial complex. It's often represented as proof that sunscreen use can prevent melanoma, but what it actually compared is exemplary use of a good broad-spectrum sunscreen verses the typical sunscreen habits of a Nambour citizen of the early 1990s, which likely meant using sunscreens that didn't effectively block UVA. It's unclear what the takeaways should be for the rest of us on the planet.

* For other skin cancers, the study found that the daily broad-spectrum sunscreen use reduced the incidence of squamous cell carcinomas by 38 percent but didn't significantly change the incidence of basal cell carcinomas, the most common skin cancers.

None of that detracts from the larger point that the broad-spectrum sunscreens that arrived in the 1990s were much better than the previous versions. And yet even with better sunscreens, melanoma rates kept climbing. They doubled again among White Americans between 1995 and 2015, from 30 per 100,000 to 60. That looks like a bona fide epidemic, and it continues to be treated as such by both the medical profession and the media.

But not everybody buys it. A surprising number of dermatologists have argued that it was mostly an illusory epidemic all along.

THE MELANOMA SKEPTIC WHO made life most difficult for his colleagues was Bernard Ackerman, founder of the International Society of Dermatopathology and the Ackerman Academy of Dermatopathology in New York City, author of seven hundred papers and sixty books, "A legend in his time," according to the *Journal of the American Academy of Dermatology*, which called him "probably the best known living dermatologist in the world" in a 2006 tribute on the occasion of his seventieth birthday.

So what did the living legend have to say about melanoma? "The field is just replete with nonsense," he told *The New York Times* in 2004, the same year that he received the 2004 Master Dermatologist Award, the Nobel Prize of dermatology. Ackerman believed that the evidence that sunscreen could prevent melanoma was weak, and that the epidemic was being driven mostly by overdiagnosis. "There has been a mania for taking off these moles that are of no consequence," he told the *Times*. "We're talking about billions and billions of dollars being spent, based on hype." He became a crusader in his later years, publishing a book in 2008 titled *The Sun and the "Epidemic" of Melanoma: Myth on Myth!*

It's one thing when an outsider says your science sucks; it's another when it comes from the guy who literally wrote your textbooks. The industry's response was to pretend it wasn't happening. His tributes tended to skirt the issue entirely, focusing on his powerful intellect, teaching prowess, and lifelong generosity. The *Times*, meanwhile, couldn't resist pointing out that he never wore sunscreen and was deeply tan.

Everyone agrees that a certain amount of overdiagnosis is unavoidable, even preferable. You want to err on the side of caution and make sure you get everything removed that might be a problem. But Ackerman was one of many dermatologists who believed that overdiagnosis had gone too far, and in recent years the field has come under increasing scrutiny.

One of the dermatologists who has looked most closely at the problem is Adewole Adamson, who directs the Melanoma and Pigmented Lesion clinic at the University of Texas. "For me it all started with a simple figure," he told me. "Melanoma incidence has gone up sixfold, and mortality has stayed flat. That never quite made sense."

Indeed, melanoma death rates stabilized in the 1980s and have since declined. At the same time, the numbers of screenings, biopsies, and diagnoses have exploded. And what those screenings mostly find are not invasive melanomas, but what are known as in situ or "stage 0" melanomas. These are small lesions that are confined to the surface of the skin and haven't yet shown signs of spreading.

Surprisingly, until a lesion becomes invasive, there's no objective way to determine if it's a melanoma or not.* It's a judgment call for

* Part of the issue is that it has turned out to be much harder than expected to determine what makes a cell go cancerous. Researchers used to think they had a handle on it because they would find certain patterns of mutations in cancerous cells, but when they sequence the DNA of normal, functional skin cells in older adults, they often find the same mutations. By the time we reach a certain age, our cells are rife with mutations, but that alone doesn't seem to be enough to guarantee cancer. Other factors are involved. Recent research points to problems with mitochondria, metabolism, and communication between cells and the surrounding body.

the pathologist. And today's pathologists are increasingly likely to diagnose new spots as melanomas. In one study, pathologists were asked to look at forty biopsies and decide if they were melanomas. On average, they called eighteen of them melanomas. But the exact same biopsies had been shown to pathologists—including some of the same pathologists—twenty years earlier. Back then, they judged only eleven of them to be melanomas.

Most of the rise in melanoma incidence has been in stage 0 melanomas. This has been spun as a success: The massive improvement in survival rates is because these baby melanomas are being caught and removed before they spread. But there are reasons to believe otherwise. The incidence rate of invasive melanomas hasn't changed much. If the early screening and removal of stage 0 melanomas was nipping real melanomas in the bud, it should have shown up in lower rates of invasive melanomas.

But invasive melanomas were appearing and acting as they always had. (The drastic improvement in survival rates happened quickly starting in 2013 with the introduction of highly successful new treatment methods such as checkpoint inhibitors, which help the immune system to find and destroy melanomas.) To Adamson, exploding cases and flat mortality looks like a classic case of overdiagnosis. "The more you look for cancer, the more you find it," he says. That leads to a runaway feedback loop as awareness expands. "Everyone in the system is playing into the cycle. Patients want to get checked. Dermatologists want to find cancer early. Dermatopathologists want to read more slides. And they're incentivized to overcall things. Nobody's going to slap you on the hand if you overcall a melanoma, but if you undercall one, you might have a lawsuit on your hands."

Every part of the system benefits from diagnosing small, enigmatic lesions as melanomas and removing them. Clinics profit. No one gets sued. And melanoma survival rates surge. "It's a vicious cycle," says Adamson. "We're manufacturing cancer survivors."

In several controversial papers in the *Journal of the American Academy of Dermatologists* and *The Journal of the American Medical Association*, Adamson said as much. He estimated that more than 85 percent of melanomas in situ are "overdiagnosed"—not true melanomas—and he suggested ending population-wide blanket screenings and raising the threshold to biopsy and to diagnose a lesion as melanoma.

Ha-ha, responded the Academy of Dermatologists. "Is it appropriate to decline to biopsy something that might kill the patient, particularly if monitoring is not an option?" the chair of the AAD's Melanoma / Skin Cancer Community Programs Committee wrote in the *Journal*. "What we are experiencing now is a swing of the screening pendulum, produced by the recognition that we may be over-screening those at low risk; we shouldn't wildly overshoot the center, though, by halting screening for those that will benefit."

On that everyone agrees—focus surveillance on the people who are likely to need it. Adamson also agrees that a fraction of the diagnosed stage 0 melanomas are real, and he concedes that the likelihood of dermatologists policing moles less aggressively is remote. "Cancer's scary! That's why this problem's intractable in many ways." Realistically, if you ask patients if they want to get their mole removed "just in case," who isn't going to say yes?

Unfortunately, the by-product of this "abundance of caution" approach was a conviction among the authorities that deadly skin cancer was lurking in every sunbeam, the public wasn't nearly as scared as it should be, and facts alone were not enough to change that.

IN 2007 DOCTORS CALLED out the American Cancer Society for running ads in a number of women's magazines showing a young woman holding up a photo of another young woman and proclaiming, "My sister accidentally killed herself. She died of skin cancer." The text urged people to use sunscreen.

The ad was objectionable because it implied that dying from skin cancer was a likely outcome (when in reality less than 1 percent of the people who get skin cancer die of it), and that if you did, it was due to your own negligence. But then it got worse when it turned out the story was made up (the women were models) and the whole campaign had been bankrolled by Neutrogena, a longtime partner of the American Cancer Society (ACS).

When confronted by *The New York Times*, the deputy chief medical officer of the ACS stuck to his guns, insisting that such shock tactics were necessary. "We have taken some license in taking that message and using it the way we've used it," he explained, "because that's the way to get the message to our target audience."

The propaganda was so egregious that the *Times* seems to have thought it necessary to go beyond pure reporting and be the grown-up in the room. It pointed out that "most skin cancer is not life threatening," questioned the practice of letting public health messaging be crafted by a company with something to sell, and quoted the associate director for disease prevention at the National Institutes of Health, who said, "There's very little evidence that sunscreens protect you against melanoma, yet you often hear that as the dominant message."

There's also little evidence that melanoma was as much of a scourge as we thought it was. While true invasive melanoma is a terrible disease to get, the rise in melanoma incidence can mostly be attributed to an aging population and shifting diagnostics. Adewole Adamson estimates that true melanoma rates for White Americans probably doubled in the forty-year period between 1975 and 2015, and some of that can likely be attributed to the inadequate sunscreens in the first half of that period that allowed massive UVA exposure.* Rates in other groups haven't changed significantly.

Today, with better sunscreens on the market and more emphasis on other sun-protective measures such as seeking shade, you'd expect to finally see declining melanoma rates—and that's exactly what we do see among younger age groups in countries from Australia to Sweden and the United States. It looks as though more use of broad-spectrum sunscreen on fair-skinned children in particular is having an effect.† The biggest factor of all may be declining use of tanning beds, which are linked to the development of melanoma in younger people—in fact, tanning beds have been shown to induce far more mutations in skin cells than natural sunlight.

With melanoma rates expected to continue to decline in coming years, we are in a good position to take a collective breath and have a more enlightened conversation about sun exposure. And there's an obvious place for that conversation to start.

IN RECENT YEARS, it has become common for health experts and media outlets alike to tell people with darker skin tones that they need to use sunscreen every day—even on cloudy days. That

* And still do. Amazingly, many sunscreens on the market still don't provide broad-spectrum coverage. Avoid them. The Environmental Working Group publishes an annual review of sunscreens that lists the pros and cons of virtually every sunscreen on the market. Opt for sunscreens on its "approved" list.

† Or maybe those kids are just staying indoors. Declining rates of melanoma go hand in hand with rising rates of myopia, which are associated with more screen time and less outdoors time.

profoundly unscientific position exposes the absurdity of the official messaging on sun exposure. There's a good debate to be had on the risks and benefits of sun exposure for people with fair skin, but for people with dark skin, there's simply no debate. The risks are minimal. The authorities know this, and it makes them nervous. They fear that admitting it will open up a large and complicated discussion about sunlight they'd prefer not to have.

Remember the team of dermatologists at the University of Manchester who exposed light-skinned volunteers (skin type II) to fifteen minutes of daily sunshine and found that it caused modest amounts of DNA damage? They also exposed people with South Asian ancestry (skin type V) to the same amount of sunlight. It caused no damage at all.

In a second experiment, the Manchester group subjected volunteers of all six skin types to increasing amounts of UV radiation up to 80 percent of the minimal amount necessary to induce a slight sunburn. This meant that the skin type IV's got three times as much radiation as the type I's, and the type VI's got seven times as much. That would have been enough to fry the lighter-skin types, and yet once again, the darker skin types showed no ill effects. Even at the highest UV doses, DNA damage in the vital basal layer was "undetectable."*

Other studies have also found that the sun-exposed skin of older individuals of European ancestry, but not African ancestry, tends

* It's worth noting that the same dose of sunshine raised vitamin D levels in dark skin only about half as much as it did in light skin, but the equivalent amount of intensity had an equal effect across skin types. In other words, a shot of UV strong enough to nearly cause a sunburn raised everybody's D levels equally, but that shot had to be seven times larger for skin type VI than for skin type I. Also of note, all the darker skin types had concerningly low levels of vitamin D, indicating that none of them were getting the sunlight they needed.

to contain mutations known to be caused by UV. Dark skin has an extraordinary ability to handle UV.

Most people assume that ability is due to the concentration of melanin, the UV-absorbing pigment responsible for tans and dark skin, but the amount of melanin is only part of the story. The melanin in dark skin comes in wider granules that migrate to the surface of the skin more successfully in response to UV light, where they act like little parasols, protecting the cells below. In light skin, the melanin tends to be scattered as "dust" throughout the skin, where it is less protective. Dark skin is also much better at repairing DNA damage, at limiting it to the upper layers of the skin, and in eliminating those damaged cells after too much UV exposure. With these taken together, it's clear that dark skin has an entire system for keeping the damaging effects of UV walled off in an outer layer of sacrificial cells.

Adewole Adamson, the University of Texas melanoma expert, specializes in treating dark-skinned patients. He told me he's spent years debunking the health-care establishment's sun recommendations for Black people. "I'd see the messaging saying, 'You're Black, you can get skin cancer, put sunscreen on!' And I just looked at the data and thought, 'This is bonkers.'"

There are a range of skin types among those who identify by any particular racial category, and from a scientific perspective it's always more useful to categorize by skin type. But most demographic data still groups people into racial categories, so that's what researchers have to work with. Even using those imperfect metrics, the results are striking.

In the United States, skin cancer is seventy times more common in Whites than in Blacks. *Sun-induced* skin cancer is estimated to be five hundred to a thousand times more common in Whites. For melanoma, incidence of new cases per year is about 28 per hundred thousand people in Whites, 5 in Hispanics, and 1 in Blacks and Asians. You find similar numbers around the world. Rates of melanoma in Asia and Africa are less than 1 per hundred thousand, and in Latin America about 4.*

The data was clear, and clearly not reflected in the messaging, so Adamson decided to go public with the information in 2019. "Many dermatology and skin-cancer-focused organizations (a few of which I'm a member) promote the public health message of sunscreen use to reduce melanoma risk among black patients," he wrote in an opinion piece in *The Washington Post*. "But this message is not supported by evidence. There exists no study that demonstrates sunscreen reduces skin cancer risk in black people. Period."

Adamson backed that up with a 2020 systematic review of all high-quality melanoma studies. There was modest evidence that sun exposure raised the risk of melanoma in fair-skinned populations, and strong evidence that it didn't in other groups. "All the metrics that show a signal in White people, the signal is absent in people of color," he says. "It's just not there. And that's okay! It should be good news. Yet somehow I'm controversial for it."

But only in less informed circles. When I asked Henry Lim, the former president of the American Academy of Dermatology, what he thought of Adamson's paper, he surprised me. "I think he's correct," he said. "His analysis is very well done. If one focuses only on people of color, there's no correlation at all between sun exposure

* On the rare occasion when dark-skinned individuals do get melanoma, it's often a type called acral melanoma, which appears on the palms, the soles of the feet, or under the nails and is not caused by sun exposure. That's the type of melanoma that killed the reggae singer Bob Marley. All racial groups seem to get acral melanoma at similar rates, but because Blacks so rarely get other types of melanoma, it makes up a significant percentage of the cases.

and melanoma. So, yes, specifically for melanoma, photoprotection is not essential for this group of individuals."*

But even for people with intermediate skin tones, it's not clear that sun exposure is a huge problem. For instance, a 2018 Italian study found that those with vitamin D deficiency (implying very little sun exposure) were twenty-five times *more likely* to contract melanoma than those with sufficient vitamin D levels. Having Mediterranean ancestry might not protect you as well as having African ancestry, but it clearly puts you in a different category from Northern Europeans.

The data on melanoma hasn't stopped the skin-care industry from pushing sunscreen for darker skin (a global market, after all, potentially six times as large as the pale-skinned one) or seeding the media with stories about the dangers of skin cancer in people of color. A particular egregious one ran on the *Today* show a few years ago, in which a young Black man who'd developed an acral melanoma on the sole of his foot was chastised for not wearing sunscreen. "He doesn't wear sunblock every day," lamented the show's go-to dermatologist. "I just had a baby cow!" Later in the segment, she drove home the official messaging: "Every day, rain or shine, January through December."

Everyone agrees that the groups crafting this messaging mean well. They are simply trying to prevent skin cancer in an underserved population. Although people with darker skin are unlikely to develop melanoma, when they are diagnosed with one, it tends to be more advanced and more dangerous, perhaps because of less ease of access to health care or because of a perception that they don't need to worry about melanoma. So the calculus in the messaging campaigns

* Lim does think that dark-skinned people should still wear sunscreen, in part to protect against sunspots. "Pigmentation is not a life-threatening issue," he says, "but it is a quality-of-life issue."

seems to be that if darker-skinned populations are induced to wear more sunscreen and get more skin checks, it might save a few lives. Where's the harm?

But a 2023 study at the University of Pittsburgh put that theory to the test and found that it didn't add up. When 12,738 people with dark skin were screened for skin cancer, only a single melanoma was found. One. "This is an almost unfathomable number of doctor's visits to find one melanoma," said the dermatologist who led the study. "Rather than screening everyone, educating physicians about presentation of melanoma in skin of color, educating the public about their risk of melanoma, and making sure that people have access to a dermatologist when they have a suspicious lesion could be more effective in improving early detection."

Adamson agrees, and he thinks any "fearmongering" that causes people with dark skin to be overly concerned about melanoma can be subtly harmful. He's seen it in his dark-skinned patients. "They get scared to go outside. They stop running or hanging out with their family in the park. Those things become unenjoyable. And there are potential health costs to that."

IT'S CURIOUS TO CONSIDER how we drifted into a quasi-religious demonization of the sun, of all things. More people die each year from choking than from melanoma, yet there's no national smoothie campaign to try to get those numbers down, no celebrity doctors on the *Today* show demonstrating how to safely eat a steak. Car accidents take five times as many lives as melanoma, yet no one checks our driving habits with the enthusiasm that they check our moles.

So why did this one public health issue grow to have such an outsize presence in the culture, to the point where *Chicago Tribune* columnist Mary Schmich could go viral with an imagined graduation speech that began, "Ladies and gentlemen of the class of '97: Wear sunscreen. If I could offer you only one tip for the future, sunscreen would be it," and the film director Baz Luhrmann could reach the top of the pop charts with his spoken-word version?

In part, as we've seen, it was just capitalism. If there was a lotion that could prevent choking, you'd be hearing a lot more about the dangers of hot dogs and cherry tomatoes. There's also a control aspect, something emblematic of our decoupling from nature. The more we retreat into the technological world of cities and screens, the more we need protection from the raw forces of the world. That's where the dragons live. Don't go without your armor.

But mostly I think we can chalk up the whole situation to random chronology. Things went down as they did because of the order in which they were discovered. If sunlight's carcinogenic effects had been discovered before its ability to cure rickets and tuberculosis and to improve other conditions, we might never have had the ultraviolet insanity of the 1920s and 1930s or the enthusiasm for tanning that stretched through the twentieth century. And if we hadn't had that, health authorities wouldn't have had to overplay their hand to get results. And if they hadn't done that, they might not have been so resistant to a reconsideration once new science began to confirm the wider benefits of sunlight.

One can easily imagine an alternative history where the light and dark sides of sun exposure are discovered together and where authorities simply present the evidence in all its uncertainty. "We don't really know" is always an excellent answer. I like to picture the celebrity derm on the *Today* show throwing up her hands and saying it's impossible to give a recommendation because individual circumstances vary so much. Instead, we get the one-size-fits-all boilerplate. No matter who you are, no matter where you are, you can never benefit from sun exposure. Stay out of the light, take your pills.

But lately that messaging has been looking shaky. New science has begun eroding its pillars. The thing about sunlight is that it shines everywhere on the planet. If it has an effect on our physiology, it's going to show up. The groups focused on skin cancer aren't going to see it, but others will.

That's what happened with sun exposure. Even as the push to shun the sun got more extreme, evidence of its benefits was building quietly in other fields. In most cases, it was because researchers pursuing something completely different stumbled upon clues that led them to sunlight. It almost had to be, since they would never have received grants to actually research sun exposure.

What they found—working first quietly, then less so—was a remarkably compelling body of information, one that tells a new story about the ways bodies respond to light and gives new illumination to some of the old scientific hunches that got derailed during the twentieth century. That story was going to force a rethink of what it means to be a living being on a photon-drenched planet. And it was going to be resisted all the way.

PART II

ENLIGHTENMENT

CHAPTER 5

CHANGES IN LATITUDE

Why Are People Less Healthy Farther from the Equator?

In the 1950s, the Minnesota epidemiologist Ancel Keys arrived in the Mediterranean with a hunch about the cause of heart disease and a pile of money to prove he was right. Keys was smart, charismatic, and imperious. By the 1950s he'd become one of the most influential scientists in the United States, and that position allowed him to launch the first in-depth international study of diet and health ever attempted: the Seven Countries Study. Keys had traveled the world with an eye to the health of its inhabitants, and he'd noticed that men in Greece and Italy seemed to experience less heart disease than men in the United States and Northern Europe. He suspected that saturated fat was the key. Northern diets included a lot of red meat and dairy, which is high in saturated fat, while Mediterranean diets got most of their fat from olive oil.

To confirm his hunch, he needed hard data, and that was what the Seven Countries Study was designed to get. From 1958 to 1964, Keys and his team recorded the diets, cholesterol levels, and heart health of 12,763 men in the United States, Finland, the Netherlands, Yugoslavia, Italy, Greece, and Japan. The results did indeed fit his theory. Men in the Mediterranean countries ate less saturated fat than men in the United States or Northern Europe, and they had lower cholesterol, lower blood pressure, and much lower rates of heart disease. Men in Japan ate the least saturated fat and had the healthiest hearts of all.

The Seven Countries Study had a massive impact on the industrialized world's eating habits. It launched the Mediterranean diet and a thousand meze platters, and most of us now follow it in some form even if we don't realize it. But as science it was weak. Keys's single-minded focus on saturated fat blinded him to other factors that might explain his results. Sugar consumption, for instance, correlated more closely with heart disease in Keys's seven countries and has turned out to be a bigger culprit worldwide.

The other glaring factor that somehow eluded Keys was environment. The clues were all around him. On his first trip to Greece in 1953, he'd remarked on the sight of "men of eighty to one hundred and more going off to work in the fields with a hoe," apparently never entertaining the idea that the hoeing in the fresh air might have been as important to their longevity as the olive oil.*

This is all the more remarkable because Keys himself was deeply impacted by the Mediterranean climate, commenting on the pleasure of leaving snowy England to drive south for his research and reaching Italy, where "the air was mild, the flowers were gay, birds were singing, and we basked at the outdoor table drinking our first espresso." Keys fell hard for the Mediterranean sunshine and spent much of his long life soaking it up in the villa he bought on the cliffs south of Naples. In *How to Eat Well and Stay Well the Mediterranean Way*, he described the scene: "Mountains behind and

* Most people don't realize that much of the work promoting the benefits of the Mediterranean diet was funded by the olive oil industry—successfully so!

the sea in front, all bathed in shimmering sunshine—that is the Mediterranean to us."

Yes, indeed, that is the Mediterranean to a lot of people. It's one of the only things the various long-lived populations of the region have in common—certainly not diet, which varies drastically from country to country. But I give Keys credit for raising the profile of epidemiology, which turns out to be one of the best ways to discover new things about the world. You look for patterns, differences between groups. And then you look for other differences that might explain what you're seeing. If you do it wrong, you get stuck on old notions. But if you do it right, you open your mind to previously unsuspected ideas. They don't always pan out, but when they do, it can turn a paradigm on its head.

ONE OF THE REASONS it's hard to fathom how Keys missed the latitude gradient in his Seven Countries data is because others had been alert to such patterns for decades. One of the first was the Columbia University physician and former New York City health commissioner Haven Emerson, who published a paper in 1932 in the *American Journal of Public Health* describing a perplexing pattern for heart disease in the United States, which was low in Atlanta and Birmingham, moderate in Richmond and Cincinnati, and sky-high in Boston and Chicago.

In 1932, the age of heliotherapy was still going strong. Emerson knew of the sun's ability to cure rickets and cutaneous tuberculosis, and he wondered if there was more to it than that. "It may well be," he posited in his paper, "that the growth and development of children and adults suffers disadvantage from insufficient sunshine or skyshine in other still unrecognized ways."

No one followed up on Emerson's suggestion, but a latitude gradient was soon spotted in other diseases, too. In cancer it came with a twist: Rates of skin cancer went up as you moved toward sunnier climes (for people with fair skin, who were the only ones to suffer significant incidence rates at any latitude), but internal cancers went in the opposite direction. That dichotomy was pointed out in *The Lancet* in 1936 by a Johns Hopkins doctor name Sigismund Peller, who noted that US Navy sailors were eight times more likely than the general population to get skin cancer, but less than half as likely to get internal cancers.

Just a few years earlier, George Marshall Findlay had induced tumors in mice with his mercury-vapor lamp, and word was getting around that ultraviolet light might be a driver of skin cancer. Peller put together the pieces all wrong, hypothesizing that skin cancer was actually protective against the formation of other cancers. He then made a bold and remarkably misguided suggestion: Since skin cancer is relatively benign and easy to treat, perhaps we should induce it in healthy people by irradiating them with UV lamps as a way of preventing the more dangerous cancers?

Fortunately, he never got to carry out his plan, and in 1940 a savvy scientist from the Medical College of Virginia named Frank Apperly came up with a better explanation. Apperly had noticed the same pattern in US states and Canadian provinces: Those receiving more solar radiation had higher rates of skin cancer mortality but lower rates of cancer mortality overall.* He proposed that the relationship was indirect: Something about sunlight itself was both causing the skin cancer and conferring protection against internal cancers. He just didn't know what.

* Intriguingly, the latitude gradient for cancer incidence isn't especially strong, but the gradient for cancer mortality is highly significant. (For example, survival time for people with pancreatic cancer is 30 percent longer in high-sun environments than in low-sun environments.) Something about sunlight either helps the immune system battle cancer, or it helps maintain the overall vitality of the body during the fight.

A candidate was proffered in 1980 by the epidemiologists Frank and Cedric Garland (the brothers who would later suggest that sunscreens might be exacerbating melanoma rates). The Garlands were analyzing maps of cancer incidence produced by the National Cancer Institute as part of the government-led War on Cancer. The maps were intended to illuminate hot spots of cancer incidence so that environmental carcinogens might be identified, but the Garlands noticed a strong north-south gradient for colon cancer rates. In a hugely influential paper in the *Journal of Epidemiology*, they suggested that vitamin D was responsible.

Until then, D was known simply as the micronutrient that prevented rickets. It was produced in the skin with the help of the UV in sunlight, and it helped deliver calcium to bones, making them hard. The Garlands suspected that it might do more than that. Indeed, everywhere scientists looked, they discovered that people with low vitamin D levels had a higher incidence of dozens of diseases, including breast cancer, colorectal cancer, high blood pressure, diabetes, heart attack, stroke, dementia, depression, and several autoimmune disorders.

By the late 1990s, a consensus had built among doctors that vitamin D was the explanation for the latitude gradient. People who received insufficient sun exposure had higher rates of disease because they were D-ficient.

You might think this explanation would have triggered a reevaluation of our faltering relationship with sunlight. If vitamin D really was a new wonder drug that could reduce rates of so many diseases, and sunlight was the main source, maybe strict abstinence was a mistake? Maybe we needed a daily dose of sunshine?

But that would have put doctors in a bind. The 1990s was the decade when the messaging on UV and skin cancer was finally starting to take hold. People were using a lot more sunscreen. They were finally beginning to avoid the sun. And this had been achieved by keeping the message exceedingly simple: The sun is your enemy. Just say no. "The sun has an incredibly complicated effect on the body" was not something Sid the Seagull was going to be able to convey in a sixty-second slot between commercials.

Fortunately, a change in message didn't seem necessary. Vitamin D isn't found in large quantities in foods, but it's cheap and easy to produce. If the only benefit of sun exposure was vitamin D, well, that could be had from a pill. No lifestyle change required. Doctors could tell patients to stay out of the sun and swallow their pills, confident that robust good health would ensue.

This advice was greeted with enthusiasm by the supplement industry, which quickly cranked out D pills by the billions,* along with accompanying ad campaigns to educate consumers about the benefits of this new wonder drug. Cancer foundations and health institutions also got in on the messaging, blanketing the media and internet with explainers debunking the "myth" that you had to get your D from the sun.

This was well-intentioned, if naive. It was convenient to believe vitamin D was the only reason people with high levels of the molecule in their bodies were healthier. It was profoundly inconvenient to think the sun might be a factor in their health, one that couldn't be reduced to a pill. That would have forced a reconsideration of all the current advice on sun exposure. So everyone just assumed D was the answer, and pills were the best way to get it. Problem solved.

* You might wonder where they found the big D-posit to mine the raw material for all those pills. It's sheep. The waxy lanolin that coats and protects their wool is rich in the same cholesterol molecules that are transformed in our skin into vitamin D. Manufacturers buy wool, isolate the lanolin, and irradiate it with ultraviolet radiation to produce vitamin D. A second variety of D is produced by irradiating yeast, which is even cheaper (and vegan), but it's less effective. That variety is known as D2, while the sheep-sourced kind is D3.

But the system of science, for all its flaws, is incredibly good at testing such assumptions. Science's very essence is its ability to ensure that we aren't just telling ourselves stories we would like to be true. And it was about to put the vitamin D story to the test and find it so thoroughly wanting that the supplement would become the poster child for failure.

BY 2010, VITAMIN D had become America's favorite health aid. "Imagine a treatment that could build bones, strengthen the immune system and lower the risks of illnesses like diabetes, heart and kidney disease, high blood pressure and cancer," *The New York Times* wrote. "Some research suggests that such a wonder treatment already exists. It's vitamin D."

The *Times* was no outlier. Everyone from health professionals to mainstream media urged Americans to get on the D wagon. "More and more studies are revealing the benefits of having plenty of D—and the dangers of having too little," declared *Oprah Magazine* that same year. "For decades vitamin D's claim to fame was its role as calcium's trusty escort, helping our bones absorb the essential mineral. But a recent flood of research is revealing that D does much more. . . . All of our bodily functions seem to rely on the nutrient, and studies show that it's key to helping prevent everything from migraines to cancer. The trouble is, most of us—53 percent of women, 41 percent of men, and 61 percent of kids—have insufficient levels."

Those sky-high rates of insufficiency were because as a society we were increasingly leading indoor lives. More screen time, less recess, longer workdays. The most straightforward solution would have been less of all that, more time under the sky, but that wasn't the direction society was heading. To fix the problem, we simply jacked up society's D levels through pills. Vitamin D became a $1.5 billion global industry, the most prescribed supplement in the world. It's still the most popular dietary supplement in the United States, taken by more than 50 million adults every day.

Most experts expected to see significant health gains from supplementation, but others worried that the recommendations were outpacing the science. All those promising signs of D's connection to lower rates of disease were coming from what are known as *observational studies*: Researchers look at datasets from large populations and notice correlations. Sometimes observational studies pan out, as they have with smoking and lung cancer, but observational studies are notorious for having *confounders*, where the differences between two groups are due to behavioral or environmental factors that haven't been anticipated.

Classic example: Those who drink coffee are more likely to have lung cancer than those who don't, but it turns out to have nothing to do with coffee. Coffee drinkers are also more likely to smoke. Once you control for smoking, the association disappears. *Correlation*, goes the famous phrase, *is not causation*.

Even the *Times* was circumspect in that 2010 article. "Don't start gobbling down vitamin D supplements just yet," it warned. "The excitement about their health potential is still far ahead of the science. . . . It may be that high doses of the nutrient don't really make people healthier, but that healthy people simply do the sorts of things that happen to raise vitamin D."

The best way to determine if an observed correlation is real is to run a randomized clinical trial (RCT), the gold standard of scientific evidence. You take two groups, randomized so they are equal in terms of health, lifestyle, and so on, and half of them get the treatment (say, vitamin D supplements) and half don't (placebo pills). Then you track them and see if any differences emerge.

To have any real diagnostic power, RCTs need to include large numbers of people and need to track those people long enough for the outcomes to emerge—which, for human health, often means years. That makes them incredibly expensive to run, meaning they get approved only for the most promising candidates. Vitamin D qualified, so it was the focus of some massive studies.

But because RCTs need to follow the participants for years, they sometimes get trapped in an unfortunate dynamic in which their results are announced long after the initial hype. First, some promising substance such as vitamin D is identified in observational studies. It all sounds exciting, media stories quickly appear, the health industry trumpets it as the next big thing, and both doctors and the public jump on it. When the sobering results of the requisite clinical trials appear years later, the horses are already out of the barn.

One of the first big tests for vitamin D was the Women's Health Initiative, which assigned 36,282 postmenopausal women either 400 IU of daily vitamin D or a placebo. The study tracked them for seven years to see if the pills would change their risk of hip fracture or colorectal cancer—both conditions that are rarer in people with high levels of D in their bodies. The supplementation showed no benefit, but the experts hedged; maybe the amount of D had been too low?

Other studies raised the dose, but it didn't make any difference. In 2014, *The Lancet* published a review of an astounding 462 studies on the vitamin. The observational studies showed overwhelmingly that people with high levels of the vitamin were healthier. And the clinical trials showed, just as overwhelmingly, that supplementing with pills did raise the amount of D in people's blood, but had no health effects whatsoever.

The final nail in the coffin was the federally funded VITAL study, which tracked 25,871 Americans for an average of five years, starting in 2011. Half the participants received 2,000 IU of daily vitamin D supplements, half received placebos. This was the ultimate test, and the results were sobering. Vitamin D supplements provided zero benefits for *any* condition, as the prestigious *New England Journal of Medicine* explained in a follow-up editorial after the final results were published in 2022: "Results of analyses from VITAL published in peer-reviewed journals have shown that vitamin D supplementation did not prevent cancer or cardiovascular disease, prevent falls, improve cognitive function, reduce atrial fibrillation, change body composition, reduce migraine frequency, improve stroke outcomes, decrease age-related macular degeneration, or reduce knee pain."

Amazingly, it didn't even prevent fractures, the one thing everyone "knew" vitamin D did. The journal pulled no punches in its recommendations: "Providers should stop screening for . . . vitamin D levels or recommending vitamin D supplements, and people should stop taking vitamin D supplements to prevent major diseases or extend life."*

Could one of the most common interventions in American medical history really be a total flop? Scientists turned to another tool of epidemiology to double-check. *Mendelian randomization* is an overly fancy name for a simple concept: Genetic variation makes some people better at producing certain compounds than others. In the case of vitamin D, a number of genetic variants have been identified that cause people to produce more or less vitamin D under the same conditions. If people who produce more D *given the same*

* There is one group to whom this advice doesn't apply: those with severe vitamin D deficiencies. For people with less than 12 ng/ml in their blood, some studies show a modest benefit from supplementation. And certainly it's well established that supplementation can prevent rickets in people who have almost no D in their bodies. So, although even for people with low levels of D the evidence of a benefit is weak, the general consensus is that for anyone with less than 16 ng/ml of vitamin D, a supplement can't hurt.

amount of sunlight are healthier, then D itself must be important. If not, then D really doesn't matter.

To answer the question, a UK team looked at the genomes of 339,000 White British individuals. They analyzed their genes to determine how good they were at making vitamin D, then checked their rates of high blood pressure, heart disease, obesity, diabetes, depression, fractures, and mortality—all conditions known to be more prevalent in people with low levels of D. In every case, their natural level of D production had no correlation to their rates of disease. D didn't matter. It was a red herring all along.

The rise and fall of vitamin D was a triumph of science, though certainly not the one the experts wanted. A theory had been raised, refined, and thoroughly smashed by rigorous clinical trials. That was exactly how science is supposed to work. But it also left the establishment in an awkward position. In trying to prove the magic of a pill, it had accidentally redeemed the sun. For decades, scientists had assumed that D was the reason people were healthier in sunnier places, so they'd never bothered to look deeper. But with D's failure in the 2010s and 2020s, they were forced to look again.

That should have been the start of a Manhattan Project for Old Sol, an all-out effort to determine if the sun really did improve health and, if so, how we could reap the rewards while minimizing the risks. But, of course, it wasn't. So much time, money, and reputation had been spent convincing people to avoid sunlight. A pivot would have been painful. Instead, the establishment did its best to ignore the issue. It *still* tells people to get their D from foods or a pill, missing the point entirely.

But, with apologies to Max Planck, scientific revolutions don't always have to happen one funeral at a time. Sometimes they happen because of the silos that everyone complains about. Maybe nobody who'd been battling skin cancer was ready to reconsider sunlight, but in all the other fields of biology, science was happening. Immunologists and cardiologists and neurologists who had no preconceived notions about sunlight—who had mostly never given it a thought either way—were hot on the trail of lifestyle and environmental factors that seemed to make people healthy. And they kept coming up with the same answer.

CHAPTER 6

THE SWEDISH PARADOX

The Mysterious Longevity of Sun-Seeking Swedes

One of the key bodies of evidence that would help convince scientists that the sun had real benefits came from an unlikely source: a Swedish project to gauge the causes of melanoma called the Melanoma in Southern Sweden study, which began in 1990. And the agent who would use that data in ways the founders of the study could never have imagined was equally unlikely: a mild-mannered obstetrician named Pelle Lindqvist.

With short-cropped salt-and-pepper hair and gentle smile lines framing his eyes, Lindqvist looks as if he were predestined to wear a lab coat and stethoscope, which he often does as a doctor and a professor at the Karolinska Institute, which awards the Nobel Prize in Physiology or Medicine. He divides his time 80/20 between patients and research. Among other things, he's an expert in venous thromboembolism—blood clots, which are five times more likely in pregnant women and those who have recently given birth. It's evolution's way of protecting against catastrophic bleeding during childbirth, but now that most babies are delivered in hospitals, bleeding is a smaller risk than blood clots, which can cause heart attacks, strokes, and other cardiovascular issues.

In the 2000s, Lindqvist was compiling a risk score to estimate pregnant women's likelihood of developing thromboembolism, based on everything from age and weight to family history and preexisting conditions. One mysterious factor seemed to be season: incidence of stroke, heart disease, and thromboembolism were all higher in Sweden in winter than in summer. As Lindqvist wrote in the paper he eventually published on the mystery, "No plausible explanation has yet been given for the seasonal variations in thrombotic complications."

But he thought he had one. Vitamin D had become a major research focus, and Lindqvist had seen some recent studies indicating that people with vitamin D deficiencies had higher levels of inflammation. He knew that Swedes' vitamin D levels tended to drop significantly in winter, due to the severe lack of sunlight. His hypothesis was that less D led to higher inflammation, and higher inflammation made blood clots more likely. Today, such an idea would be unsurprising—inflammation is a factor in many diseases, and it will play a large role in this book—but twenty years ago, few people were paying attention to it.

Frustratingly, Lindqvist had no way to test the idea. Initiating a study of sun exposure and seasonality and its effect on thromboembolism or anything else would be a major undertaking. Epidemiology is only effective when it has the statistical power that comes with huge datasets. You need lots of participants and extensive questionnaires, and if you're trying to measure a disease that can take decades to manifest, you need to follow your participants for many years. The time, money, and hours needed can be prohibitive.

Lindqvist had no such dataset, and no expectation of being able to produce one, so he tucked away his hypothesis about vitamin D and blood clots and continued with his normal work, until a lucky break changed the course of his life's research.

LIKE MANY A SWEDE, Pelle Lindqvist's strategy for thriving in a challenging northern climate is to throw himself into it. Cross-country skiing, saunas, a deep love of the silence of snowy woods. Every Sunday in winter, he meets a dozen friends on the frozen Baltic Sea and they wild skate for hours. In the fall, he goes deer hunting, and that was when his breakthrough unexpectedly struck.

He was hunting with a buddy from Lund University, an oncologist named Håkan Olsson, idly describing his theory and his frustration at not being able to pursue it. "Håkan just laughed," Lindqvist says. "He wasn't convinced. But he's a real researcher, with an open mind." The more Lindqvist sketched out the threads of evidence, the more curious Olsson became. And he told Lindqvist he knew exactly how to test the idea.

In 1990, the oncologist had helped launch the Melanoma in Southern Sweden study, which interviewed forty thousand Swedish women about every detail of their health and lifestyle habits, then continued tracking their well-being, reinterviewing them periodically. The goal of the study had been to establish risk factors for melanoma, which is more prevalent in Sweden than anywhere else in the world except for Australia and New Zealand.*

In addition to all the usual questions about diet, income, education, exercise, family histories, and so on, the study asked multiple questions about sun exposure. How often did the women sunbathe

* After Australia and New Zealand, the countries with the next-highest rates for melanoma are found in Scandinavia and Northern Europe, a clear sign that the biggest risk factor for melanoma is fair skin.

in summer? In winter? Did they work outside? Did they go south on vacations and sunbathe? Did they ever use tanning beds? The purpose was to assess how much various "sun-seeking behaviors" increased the risk of melanoma—but Lindqvist realized that the same data was gold for investigating other effects of sun exposure as well.

Lindqvist excitedly teamed up with Olsson, acquired the data, and ran the numbers. The results blew him away. Sun-seeking Swedish women had dramatically lower rates of thromboembolism. And the population as a whole had a 50 percent higher risk of blood clots during the winter. The dataset also showed that smoking raised the risk by 40 percent, and obesity by 150 percent, while strenuous exercise cut it in half. Since those were all well-documented factors one would expect to see, it gave Lindqvist more confidence that the data was solid and the numbers on sun exposure weren't an aberration.

Lindqvist and Olsson chalked up the effect to vitamin D and published the findings in the *Journal of Thrombosis and Haemostasis*, where they made zero splash in the world.* But Lindqvist and Olsson were just getting started. They had a spectacular dataset tracking the health and sun habits of forty thousand Swedes and a lot more questions to ask. What about diabetes? That was another blood condition associated with low vitamin D rates. Could sunlight be a factor?

Once again, the correlation was striking. Sun lovers had the lowest rate of diabetes, moderate sun seekers were 47 percent more likely to get the disease, and sun avoiders were a startling 147 percent

* Except in Sweden, where Lindqvist is responsible for crafting the Swedish government's guidelines on preventing thromboembolism. He updated the guidelines to recommend that pregnant women at high risk of thromboembolism receive information about sun exposure—the only country in the world with such a recommendation on the books.

more likely. Scientists are more convinced by data when it shows that kind of "dose-dependent" relationship, where a small dose of the thing being measured has a small effect, and a large dose has a bigger one.

More confirmation: Smoking raised the risk of diabetes by 30 percent and exercise dropped it by 40. The data was good, but also eye-opening. As Lindqvist put it to me, sun exposure seemed to be about as good for you as exercise.

Lindqvist and Olsson noted that blood-sugar levels tended to be higher in winter, pointed the finger at vitamin D deficiency once again, and published the results in *Diabetes Research and Clinical Practice.* The world went on as before.

But in 2011, having established that sun exposure seemed to strongly improve two different cardiovascular conditions, and knowing that cardiovascular disease is the leading cause of death worldwide, Lindqvist and Olsson went hunting for big game in their next analysis: all-cause mortality.

After having parsed the data every which way for their previous papers, they were pretty sure they knew what they were going to find, but the magnitude of the results was still alarming. Over the twenty years that the women had been tracked since 1990, sun avoiders were twice as likely to die from any cause as sun lovers. Again, the relationship was dose dependent: Those who practiced some but not all of the sun-seeking behaviors were 40 percent more likely to die than the sun lovers, but much better off than the sun avoiders.

In one of their papers reporting the results, Lindqvist and his coauthors made an important point: The cohort in the study consisted of one thousand women born each year from 1925 to 1965, an era when Sweden experienced little immigration, so they were an extremely homogeneous group of light-skinned ethnic Swedes adapted to the low-light environment. "If avoidance of sun exposure is a major risk factor for all-cause mortality in the case of Caucasian women," they wrote, "the problem may even be more serious amongst women who traditionally cover their skin or women more densely pigmented." In other words, sun deprivation could be a massive, and massively overlooked, health emergency for the large portion of the global population that doesn't have fair skin.

This time, the paper scored a prominent spot in the 150-year-old *Journal of Internal Medicine* and was made even more prominent by an accompanying commentary from Penn State's Nina Jablonski, by then well established as a leading thinker on matters of the skin. Titled "Is There a Golden Mean for Sun Exposure?," the editorial explained the background of the Melanoma in Southern Sweden study and noted the irony that such revelatory findings were coming from a study intended to prove the dangers of sun exposure. "At the inception of the study in 1990, the investigators almost certainly did not anticipate that avoidance of sun exposure would be associated with an increase in all-cause mortality, but that is exactly what they discovered."

Jablonski then homed in on one of the most tantalizing points the paper had raised. If sunbathing in winter (in Sweden) was effective at reducing mortality, that couldn't be because of vitamin D. Only UVB light is capable of producing vitamin D in the skin, and in high-latitude places such as Sweden, almost no UVB makes it through the atmosphere from September through April. So Lindqvist's study, she wrote, "supports the argument that some unidentified factor correlated with sun exposure is protective. . . . The mortality-reducing effects observed amongst the most active exposers probably were due to many factors including a suite of bioactive compounds produced by both UVB and UVA exposure."

Jablonski called for more research to determine what these beneficial compounds might be, and what happens to people who don't get enough of them, and ended by asking, "What is a modern human supposed to do?"

She then answered her own question. A lifestyle lived almost entirely indoors was vastly different from what we had evolved for and was likely doing us no favors. In such an unnatural setting, she suggested that we need "to compensate for modernity through bespoke prescriptions for sun exposure and diet that are appropriate to our ancestry, location and lifestyle."

As for the search for those unidentified factors, the timing of Jablonski's editorial couldn't have been better. One of the most important, and unsuspected, had just had its coming-out party.

CHAPTER 7

RELAX

A New Explanation for the Sun's Cardioprotective Powers

Nitric oxide is a common gas that we now know plays many important roles in the body. That discovery earned the pharmacologist Robert Furchgott a Nobel Prize in 1998. It would be another decade before others figured out that nitric oxide was also the secret to some of sunlight's magic, but the sun had actually dropped a big hint way back in 1953, when a few rays of its light fell upon a strip of rabbit aorta floating in a jar in the window of a young Furchgott's Washington University lab.

Furchgott was investigating the ability of various drugs to make blood vessels dilate and contract, in hopes of identifying future medicines that could improve blood flow. For his experiments he used strips of rabbit aorta, which he would combine in test tubes with various compounds. Most of that work took place under the pallid light of a 1950s lab, but on that day in 1953, Furchgott happened to leave one of his jars on the windowsill, and as the sun fell upon it, he watched in amazement as the vessel relaxed, dilating visibly. A cloud passed over the sun, and the blood vessel contracted. Then more sun, and it dilated again.

Furchgott had no idea why sunlight would cause a blood vessel to dilate. He called the effect photorelaxation, but he never had much time to pursue it. He was too preoccupied by his main research, which focused on identifying the mechanism that caused blood vessels to dilate inside the body. Obviously, sunlight was not a factor. But if he could figure out which signaling molecule the body used to make the smooth muscle of the blood vessels relax, that could lead to a whole new class of drugs to improve constricted blood flow, reduce blood pressure, and prevent heart disease and stroke.

It took more than thirty years for Furchgott to discover that the mysterious substance was nitric oxide, which he announced at a landmark 1986 symposium. It had been so hard to find because nitric oxide is so volatile, disappearing from the body seconds after being produced.

The revelation shocked the medical world and transformed medicine. Until then, most scientists believed that all the chemical messengers and transmitters in the body were long molecules with complex 3D shapes that locked into matching receptors on cells. It had never occurred to anyone that a molecule as simple as a single nitrogen and a single oxygen joined together could be so key to the functioning of the body, but it actually made sense. The tiny size of the molecule and the volatile nature of the gas meant that when produced by specialized enzymes in the lining of the blood vessels, it would instantly diffuse through the surrounding muscle cells, telling them to relax.

The announcement triggered a gold rush as researchers raced to find out everything they could about nitric oxide, which turned out to have an astonishing multitude of uses in the body. Not only does it dilate blood vessels—which has led to a wave of new cardio drugs, as well as Viagra—it's also used by the brain to transmit signals, and by the immune system to kill pathogens. It even improves the efficiency of mitochondria, the tiny structures in our cells that transform nutrients into usable energy.

On the day in 1998 that the BBC World Service announced Furchgott's Nobel Prize, one of the people listening to the radio broadcast was a young nitric oxide researcher and climbing nut named Richard Weller, who was huddled inside his tent high in the Himalayas on a subzero night, near the base camp of a mountain called Ramtang. When the crackly announcement came through on his shortwave radio, Weller screamed and leaped out of his tent, dancing in the snow in his boxer shorts and alarming his climbing mates. His field of research had just been coronated as the hottest thing in medicine. And he was soon to push it forward and cause a stir of his own.

"I'M NOT BY NATURE A REBEL," Richard Weller insisted when I first chatted with him about his controversial research back in 2018. "I was always the good boy that toed the line at school. This pathway is one which came from following the data rather than a desire to overturn applecarts."

After an English childhood, the good schoolboy had grown into a long, lanky medical student with a serious climbing addiction. He spent every summer in the Alps, tackling TD or *très difficile* routes—not the north face of the Eiger, but not far from it. When it came time to choose a residency, he headed for Scotland because its mountains were better than the English ones.

Once there, he couldn't help but notice the huge discrepancies in health between people living in his new home and those in his old one in southern England. Life expectancy in Scotland was more than two years lower than in England. The primary culprits were higher rates of heart attack and stroke, far more than could be accounted for by the higher smoking rates up north.

Weller knew that blood pressure was higher in winter, and that Scots were 30 percent more likely to die in winter than in summer. Curious stuff, but far from his area of expertise, so he just tucked the information into the back of his mind. He could never have anticipated that his research was going to lead him straight back to it.

While doing his dermatological rounds in the psoriasis ward at the University of Aberdeen, Weller bumped into a pharmacologist named Nigel "Ben" Benjamin, who was deeply interested in nitric oxide. It was the 1990s, Furchgott hadn't yet won his Nobel, but nitric oxide's ability to relax blood vessels and deliver more blood and oxygen to tissues had already shown promise for improving athletic performance and treating everything from hypertension to erectile dysfunction, and it was the focus of intense research.

Benjamin suspected it might be a mediator of inflammation in the body. One of the characteristics of inflammation is an increase of blood flow to the affected area, as the body rushes immune cells to the site to attack the perceived danger. Since nitric oxide increases blood flow, the idea made some sense. And if it did indeed mediate inflammation, it might play a role in psoriasis, an autoimmune disorder in which the body's immune system attacks its own skin cells, producing inflammation and itchy, scaly red skin.

Did that mean psoriasis patients would produce nitric oxide in their skin? To find out, Benjamin needed a patient. Weller recruited one of his psoriasis patients, shrouded the man's arm in a plastic bag, and shoved a chemiluminescence probe inside. Sure enough, the man had nitric oxide emanating from his arm. Weller and Benjamin were elated. They'd just discovered a whole new phenomenon, and a possible new approach to treating psoriasis.

"And then Ben said we need a healthy control," Weller recalls. "So I shoved my arm in."

If there was no nitric oxide coming off Weller's arm, they'd have a strong sign that nitric oxide was involved with psoriasis. But to their surprise, Weller's arm was off-gassing, too. They hadn't discovered something new about psoriasis after all. After a moment's disappointment, however, they realized they'd actually discovered something more important: Apparently *everybody's* skin produced nitric oxide. They had no idea if that had any biological importance, but for now it didn't matter. The discovery of its existence was enough.

"That half hour changed my life," Weller says. Over the next six months, squeezing in extra lab time after hours, they figured out what was happening. The skin contains stores of nitrate, which is formed by a nitrogen atom with three oxygen atoms attached. It's a common compound found in vegetables, especially leafy green ones.* When we sweat, the combination of bacteria and acid in the sweat breaks down nitrate into nitric oxide, which instantly escapes into the air.

"After work, we'd go over to Ben's house for dinner and then, after dinner, sit down at his computer and bash out the paper," Weller says. "Sent it off to the *Journal of Investigative Dermatology*, the top journal. Instantly accepted, of course, because nitric oxide. And I thought, 'This is fun! This is easy!'"

As he recounts this story, Weller chuckles darkly at the naivety of his younger self. "Well, nothing has been as easy since."

THE NEXT FEW YEARS were "horrible," says Weller. He determined that the skin is loaded with nitrogen compounds, way more

* Which means, yes, greens and root vegetables such as beets can boost your nitrate supply, increase blood flow, and enhance athletic performance. Eat your greens.

than are found in the bloodstream, but he couldn't figure out what they were doing there. At first, he'd suspected that nitric oxide might play some role in skin cancer, but all those leads had been dead ends.

The sweat process he'd discovered also didn't seem terribly important. Perhaps a small amount of nitrates was feeding the skin microbiome, but that couldn't account for the "shedloads" of nitrates he was finding in the skin. Nothing could. They were being used for something, but he couldn't figure out how or why, and because of that, he couldn't publish. He felt his scientific career sputtering. All he could do was brood on the mysteries. *Why* was the body going to so much trouble to park all those nitrates in the skin?

As has so often been the case in Weller's life, revelation came in the mountains. This time, he wasn't climbing; he was helping to run the World Nitric Oxide Conference, which was being held at the opera house in Bregenz, Austria, high in the mountains near the Swiss border.

And as is so often the case at scientific conferences, the breakthrough came not during the day's presentations but in the bar afterward. In this case, it was not just any bar, but a dramatic one, overlooking the floating stage where a famous sequence in the Bond film *Quantum of Solace* had just been filmed.

As Weller shared a pint with his fellow nitric oxide nerds, Lake Constance sparkling jewellike behind them, they swapped research notes. Weller mentioned that he'd discovered that the skin was loaded with nitrates, but he didn't know why. After a moment, his German colleague Martin Feelisch spoke up. He had coauthored a paper showing that ultraviolet light could break nitrate into nitric oxide. The goal of the paper had been to show why attempts to measure the amount of nitric oxide in the body can be thrown off by light, but Feelisch hadn't explored the biological significance, and few people had read his paper.

Weller could feel the gears of his mind locking into place. "It was one of those eureka moments!" he says. "Right there in the bar!"

The reason the body was shunting all those nitrates from food into the skin was so it could take advantage of the sun's ability to cleave them into nitric oxide. In addition to its vasodilating powers, nitric oxide is a powerful antioxidant that can protect against sun damage. It can also boost wound healing by recruiting more immune cells to the site.* It all suddenly looked like a clever system, where the more sun exposure one got, the more nitric oxide would be generated in the skin to heal the damage.

But there was more to it than that. If nitric oxide produced in the skin also dispersed into the bloodstream through the capillaries in the skin's lower layers, the way vitamin D does, it could dilate blood vessels and lower blood pressure. The sun would be doing the same thing the latest nitric oxide–based hypertension drugs were doing. And that would explain why people in sunnier places had lower hypertension and mortality.†

The implications were immense, and not just in terms of potential new treatments. Weller couldn't help but see the other side of the equation, too: By urging everyone to avoid sun exposure at all

* This is probably why those surgeons in World War I noticed that wounds healed better in sunlight.

† This also helps us make sense of the lower blood pressure and heart-disease incidence Ancel Keys saw in Mediterranean countries in his early studies. People in the Mediterranean were eating a lot more leafy greens and other veggies than people in Northern Europe, which would have led to higher nitrate stores in their skin—but that alone wouldn't have been enough to lower blood pressure and protect their hearts. They also needed ample sunlight to liberate all that nitric oxide from storage. In confirmation of this idea, simply adopting a Mediterranean diet has been shown to lower blood pressure, but not as much as increasing sun exposure does. It's the combination of the two that does the trick and should be the foundation of a true Mediterranean diet.

costs, had the world's authorities inadvertently exacerbated cardiovascular disease?

WELLER AND A FEW other big names from the conference raced off and banged out what Weller calls their "lamppost papers," marking the territory as their own. Weller's first appeared in the *Journal of Investigative Dermatology* in 2009 and focused on his measurements of the huge stores of nitrogen compounds in the skin.

Having established that, he made some revolutionary suggestions about what it meant. He noted that the amount varied greatly from person to person and likely had to do with how many greens and root vegetables they were eating. He then proposed that sunlight could liberate the nitric oxide in these stores, which would not only protect the skin from ultraviolet damage but might also get into the circulation and help to relax blood vessels.

His second paper, written with Martin Feelisch in 2010, focused on the big picture. Provocatively titled "Is Sunlight Good for Our Heart?," it acknowledged that sunlight raises the risk of skin cancer, but proposed that it also liberated lots of nitric oxide in the skin, which then dilated blood vessels, and that was responsible for the lower blood pressure and mortality in sunnier places and seasons.

Having proposed it, Weller then set out to test it. Back in his Edinburgh lab, he took twenty-four medical students and shined UVA light on their arms at an intensity equivalent to thirty minutes of midday sun in Southern Europe. The use of UVA instead of UVB was important. At the time, people still assumed that any suggested benefit of sunlight was due to vitamin D, which is produced by UVB but not UVA. If the UVA light lowered blood pressure, it couldn't be because of vitamin D.

Sure enough, Weller's subjects' diastolic blood pressure dropped an average of about five points. That result matched up almost perfectly with a similar study one of his colleagues from the Bregenz conference had done in Germany: UVA really did lower blood pressure.*

The drop wasn't huge, but it had come from just that small dose of light, so it was likely that regular sun exposure would have a much bigger effect. And as Weller pointed out in his paper about the experiment, even a reduction of five points would be enough to reduce the risk of stroke by 34 percent and heart disease by 21 percent—and because these are two of the leading killers in the world, on a global scale that could save millions of lives.

It was the most important finding of his life, and it scored him another paper in the *Journal of Investigative Dermatology*. It also shed new light on the cardioprotective benefits of sun exposure seen in Pelle Lindqvist's Swedish studies—nitric oxide, not vitamin D, was the key all along.†

Weller felt that this information should be spread beyond the usual specialists, though he hadn't a clue how to do that. Fortunately, however, a new organization that thrived on overturning applecarts was coming to town. In March 2012, a local TED event was held in Glasgow for the first time, and someone who knew Weller was a natural storyteller asked him to present. He quickly threw together a twelve-minute talk titled "Could the Sun Be Good for Your Heart?"

He opened with mysteries: Australians have a third fewer deaths from heart disease and stroke than Brits, while Scots have 23 percent

* In a later study, Weller checked a three-year record of the blood pressure of more than 342,000 patients in 2,178 dialysis centers in the United States, who had their blood pressure taken multiple times per week, and compared it to NASA satellite data showing exactly how much sunlight had been falling over each dialysis center at the time. The more sun, the lower the patients' blood pressure.

† Since then the effect has repeatedly been confirmed, most recently by the UK Health Security Agency, the government agency tasked with protecting the public health, which found that a small dose of English summer sun was enough to trigger a significant release of nitric oxide that went on for days after the exposure.

higher mortality rates than the English. What could account for that? Vitamin D? No, because raising the level through supplements doesn't change the rate of heart disease.

Well, how about sunlight? Weller took the audience through his experiment shining UV on volunteers, producing nitric oxide, and lowering their blood pressure, with no change in their D levels. He showed how this could be a major factor in heart disease worldwide, and he ended with a classically provocative TED moment: "I mean, I'm a dermatologist. My day job is saying to people, 'You've got skin cancer, it's caused by sunlight, don't go in the sun.' I actually think a far more important message is that there are benefits as well as risks to sunlight. Yes, sunlight is the major alterable risk factor for skin cancer, but deaths from heart disease are a hundred times higher than deaths from skin cancer. And I think we need to find the risk-benefit ratio. How much sunlight is safe, and how can we finesse this best for our general health?"

It was a question few dermatologists had ever proposed, certainly not in such a public forum, and it caught fire. It immediately changed the calculus for me, in part because its simplicity and logic were so sound. If modest sun exposure lowered blood pressure and improved cardio health, that *had* to be part of the conversation, and sun deprivation needed to be listed alongside poor diet, lack of exercise, and smoking as a cause of heart disease, stroke, diabetes, and other cardiovascular illnesses in the modern world.

Weller's talk was a hit, and it was quickly picked up by the main TED website. By the time it had been viewed more than a million times, Weller had transformed from good soldier to rebel commander. Briefly, he says, he felt that he was all alone out there, challenging an empire with nothing but his one experiment and a few theories. "And that's when this mysterious Swedish obstetrician came along."

CHAPTER 8

ALL CAUSE

The Case for Sunlight Comes Together

The year 2014 marked an inflection point in the sunlight story. Weller was reading Pelle Lindqvist's work on sun exposure in Swedish women. Lindqvist was reading Weller. They were both discovering the work of other researchers who had for years quietly been questioning the conventional wisdom on sun exposure. Now they were finding one another and discovering that it all added up.

"There's now quite a big circle of us in the sunlight-is-good-for-you world," Weller told me. "We're out in the open, writing papers. The data is getting stronger and stronger." In Australia, researchers were shining UV light on mice and finding that it protected them from diabetes, obesity, and other metabolic disorders. It was only mice, but it was one more star in the constellation of evidence. In the UK and the Netherlands, a study of more than ten thousand people found that levels of insulin resistance and metabolic dysfunction were much lower on sunnier days. Something about sunlight seemed to improve the way energy flowed through the body.

Emboldened, Lindqvist and his team published a follow-up study noting that over the twenty years since the original Melanoma in

Southern Sweden study, nonsmokers who avoided sun had similar rates of death to smokers who sought sun. (Smokers who avoided the sun had the worst rates of all.) They concluded with the now-immortal words "Avoidance of sun exposure seems to be a risk factor of magnitude similar to smoking in terms of life expectancy."

Well. That was the last straw for the American Academy of Dermatology. Its president fired off a letter to the *Journal of Internal Medicine* stating the academy's objections. "On behalf of the 18,000 members of the American Academy of Dermatology, I am writing to express concern," it opened, adding, "We believe the results of this study may encourage the public, women in particular, to seek the sun and/or use indoor tanning devices, ultimately increasing their risk for skin cancer."

Then it played the "correlation is not causation" card, which scientists tend to do anytime they don't like the results of some observational study. Sure, the people getting more sunlight were healthier, but that didn't prove the light was responsible. Something else about that sunnier environment must be the factor. It could even be *reverse causation*—maybe the women staying inside were already less healthy in some way.

The AAD letter pointed out that you can't tell much about someone's behavior from a few questions in a study, and anyway, the Swedish sun is weak, so the results don't apply to sunnier regions. Most concerning, the president wrote, "The study's suggestion that an increased risk of skin cancer is acceptable minimizes the morbidity associated with these diagnoses and overlooks the associated economic burden of treatment." The letter reiterated the AAD's standard position: Avoid the sun, get your D from diet or a pill.

Lindqvist and Olsson fired back: "It is correct that those following the AAD's guidelines with 'as little sun exposure as possible' will have few cases of skin cancer. However, according to our observational findings, it might also shorten their lives by 0.5–2 years. We believe that those giving such advice take a lot of responsibility."

The two pointed out that melanoma was the only type of skin cancer associated with increased mortality and cited research linking it to intermittent sunburns but not daily exposure. They also pointed out that the women in their study diagnosed with nonmelanoma skin cancers also had the highest life expectancy. In Sweden, where vitamin D deficiency was rampant, perhaps this was a good trade-off? And as for the sun-seeking women being different in some invisible way, the two groups had been matched in terms of exercise, diet, BMI, smoking, education, income, and all the usual suspects, so it would have to be quite a mysterious confounder indeed.

There were more letters, along with a flurry of articles in the popular press. Naysayers kept arguing that there was no way to prove that sun seekers weren't already healthier than sun avoiders.

But there was.

THE ACID TEST OF any scientific experiment is that it must be reproducible. If you've really discovered something true about the world, others should be able to get similar results from similar experiments. To see if Pelle Lindqvist's controversial idea that sun seekers lived longer held true in the UK as well, Richard Weller teamed up with the epidemiologist Chris Dibben. They went straight to the UK Biobank, that massive trove of data on hundreds of thousands of Brits.

One of the many characteristics tracked in the UK Biobank data was whether people used sun beds. People who do use them not only get UV exposure from the beds, they also tend to seek out more sunlight in general, as previous studies had shown. So sun-bed use is a good indication of sun-seeking behavior. Dibben and Weller found that those sun bedders were 15 percent less likely to die than sun avoiders. That was extremely close to Pelle Lindqvist's findings, and it was after adjusting for all the standard confounding factors.

In theory, one could still argue that the sun seekers were just *different* from the sun avoiders in some healthy way that had nothing to do with exercise, smoking, income, and so on. So Dibben and Weller then checked the mortality figures in the same dataset against geographic location. Those in southern England were 12 percent less likely to expire than their mates in foggy Scotland.

Importantly, the healthier populations in each half of the study were very different from each other. The tanning-bed users were on average younger, less educated, and more likely to smoke than the overall population, while the southern Brits were older, more educated, and less likely to smoke. So the odds that the improved longevity in the two groups was due to some cryptic behavior that they had in common was remote.* The effect was real. Once again, the biggest drop was in cardiovascular deaths, but there was a drop in mortality from all the major cancers as well.†

With those numbers in hand, Weller got curious about how many people in the UK Biobank dataset had died from too much or too little sunlight. When he and his graduate student ran the numbers, they found that in the fifteen years of tracking, a total of

* Dibben also incorporated a "negative control"—car accidents. This is an epidemiological strategy to make sure you aren't unwittingly selecting for some difference in behavior in your groups. Car accidents shouldn't correlate with sun exposure. But if, hypothetically, sun-bed users had a higher rate of accidents than nonusers, that might tell you that they behaved differently as a group in some way that might be skewing other results. In this case, car accidents showed a similar prevalence in all groups.

† In 2025, using a powerful new tool called Olink, which can analyze the exact proteins present in a single cell, Weller and the health start-up Cytokind discovered a distinctive suite of immune molecules present in sun-exposed people that may well be responsible for this anticancer effect. As this book went to press, those results hadn't yet been revealed.

40 people had died from skin cancer attributable to too much UV light, while 2,982 people had died from diseases attributable to a deficiency of sunlight. About a seventy-five-to-one ratio—and a bad trade all around. "We have got our health message catastrophically wrong," Weller says.*

AS GOOD AS THESE studies are, they all measure sun exposure indirectly—through satellite data pegged to the participants' zip codes, or self-reporting of activities. That's not ideal. Neither one is a direct and objective measure of how much sun exposure an individual received. To get that, you need to run experiments in which you attach light monitors to volunteers. Fortunately, wrist-watches that record light levels are now readily accessible, and that has resulted in some recent high-quality data on light and health.

A 2020 Brazilian study fitted 103 subjects in one town with wrist monitors that tracked both light and motor activity. There was a strong association between light exposure and metabolic syndrome, the cluster of conditions including insulin resistance, glucose intolerance, high blood pressure, and obesity that lead to diabetes, heart disease, and the whole suite of cardio problems. Those who received more light during the day had fewer metabolic problems, as did those who were exposed to less artificial light at night. It had nothing to do with activity or amount of sleep, which was similar for the two groups—a surprise to the researchers, who concluded, "Adequate sun exposure appears to be important for cardiometabolic health."

That is almost identical to the conclusions of the giant 2024

* Intriguingly, Weller also found that high-sun Brits had almost a 60 percent reduction in tuberculosis incidence compared to low-sun Brits—a hint that the heliotherapy crowd of the early 1900s may have been on the right track all along.

UK Biobank study I mentioned in the opening of the book, which tracked eighty-nine thousand volunteers with light sensors and found that, over the following eight years, the highest-daylight-exposure group was 17 percent less likely to die from any cause. "Personal day light exposure appears to be an independent predictor of mortality risk," wrote the researchers.

As with the Swedish and Scottish studies, the biggest benefit was in cardiometabolic mortality. So clearly nitric oxide is an important part of the explanation. Sun hits skin and releases nitric oxide, which dilates blood vessels, lowers blood pressure, and saves lives.

But across these studies, the effect on other causes of mortality was almost as large, and that can't all be explained by blood pressure. The Case of the Salubrious Sunlight grows more complicated. The breadth of benefits seems to indicate that sunlight operates on a deeper and more profound level, keeping the body on an even keel. To understand how, and why, we have to think more holistically. What keeps a body sailing along through the many storms life throws at it? As it turns out, the answer to that question, like so many other holistic questions about health, leads straight to the immune system.

CHAPTER 9

FRIENDLY FIRE

Modernity, Chronic Inflammation, and the Sun Deficit

Of all the jobs held by all the parts of your body, the immune system might have the hardest. Every day, it patrols your body from the skin to the deepest recesses, trying to identify what's a healthy part of you, what's a harmless piece of the world that just happens to be passing through, and what's an invading microbe, a cell that's turned cancerous, or some other unwanted exotic. When it finds something suspect, it calls in the big guns, eliminating the problem with a variety of molecular weapons. To its credit, it makes few errors. But occasionally, it identifies some of the body's own cells as problems and trains its fire on them. The result is an autoimmune disease.

Many autoimmune diseases seem to be associated with sun exposure, but the one that leaps out is multiple sclerosis, or MS, a terrible disorder in which the immune system mistakenly attacks myelin, a fatty material that coats nerves like the plastic covering that coats the wiring in your house, preventing the electrical signals from shorting out. As immune cells destroy the insulating myelin, brain signals stop transmitting, leading to a host of physical and neurological breakdowns.

Scientists have always been astonished and perplexed by the steepness of the MS latitude gradient. Incidence rates are close to zero near the equator and then steadily rise an average of 3.6 cases per 100,000 people for every degree of latitude, reaching about 200 cases per 100,000 people in such places as Canada, the UK, and Northern Europe.

The connection had already been spotted a century ago, when doctors noticed that US servicemen from the northern states were more likely to have MS than those from the south. All kinds of causes were floated—climate, diet, cosmic rays, something in the water—but a 1960 study of veterans showed that incidence of MS tracked closely with average annual sun exposure, and the steepness of the gradient was startling. On the West Coast, for example, it was seven times as common in Seattle as in Los Angeles, with San Francisco in between. The authors asserted, "The more sunshine there is in a climate the less multiple sclerosis there appears to be." They didn't pretend to understand why, but the association was so strong that MS came to be nicknamed the sunshine disease.

Until recently experts still had no idea how sunlight could be a factor in an autoimmune disease. That's now changed, thanks to some remarkable and little-known research, and it's why I've come to think of MS as the Rosetta stone for deciphering how light translates into health.

Up to now, the possible explanations for sunlight's effects have been like those for vitamins. Light produces some *thing* the body needs, whether vitamin D or nitric oxide or some other molecule, and when you don't get it, you have a deficiency that causes a problem.

That all seems to be true. But that kind of deficiency seems less likely to explain any miscalculations of the immune system. For that, we need to think holistically and look more deeply into what a body is and how it navigates the flux of daily existence. The more you look, the more astonishing it seems that we make it through a single day, much less a near century. And the more questions you ask about what we can do to make the next day go well.

ONE OF THE SCIENTISTS who has closely examined the links between sunlight and MS is Robyn Lucas, an Australian epidemiologist who studies the impact of environment on disease. Some of the best data come from Lucas's home country, which is one of the few to boast a wide range of latitudes, a relatively homogeneous population, and a national health-care system with excellent recordkeeping and relatively consistent levels of care. Despite being a generally sun-drenched country, Australia still has MS rates six times higher in its high-latitude southern regions than its sunny northern ones, and there's a consistent gradient in between.

In the early 2000s, Lucas was part of a team studying environmental factors that might contribute to MS. They looked at such things as smoking and exposure to Epstein-Barr virus, but latitude jumped out at them. When she parsed the data, she found a mild correlation between MS and vitamin D, but "I actually found a much stronger effect for sun exposure." And not just in Australia. It was true in New Zealand, Europe, the United States, pretty much everywhere she looked. It showed up in other measures of sun exposure, too. Those with the most sun damage on their skin had just one-third the rate of MS compared to those with less. Even within the same region, MS relapse rates followed a seasonal cycle, higher in winter, when sun is scarce.

The conditions that lead to MS are believed to be set during gestation and childhood, when the young brain and immune system are first developing. There, too, sun exposure seems to play a role. Incidence rates correspond to month of birth, being highest in cases where the first trimester of pregnancy took place during winter. The pattern also shows up when you measure kids' outdoor time. Kids who spend less than thirty minutes a day outside have twice the risk of MS compared to those who spend up to one hour outside, and about five times the risk of those who average more than an hour outside.

These observational studies have been backed up by a handful of small pilot studies testing narrowband, noncarcinogenic UV light as an MS therapy. (Scientists are skittish about using full-spectrum sunlight in trials, and foundations are unlikely to fund such a trial anyway.) The most interesting study was by Prue Hart, an immunologist at Australia's Kids Research Institute, who has been studying the effects of sunlight on immunity for more than thirty years. She recruited twenty patients with clinically isolated syndrome, an early-stage version of MS that leads to full MS. Half the subjects received eight weeks of narrowband UV, a few minutes per session, three sessions per week. The other half got a placebo light.

Three months into the trial, the placebo group's disease severity had increased, while the UV group's had lessened. The UV group reported less fatigue. A year after the start, all of the placebo subjects had developed full-blown MS, but 30 percent of the UV group had been spared. Importantly, the effect lasted for months after the initial treatment.

Nothing in science is ever 100 percent certain, but for the "sunshine disease," Robyn Lucas believes the entirety of evidence is so consistent across so many measures that it has overcome her professional skepticism. "Is the evidence in harmony? Does it all hang together? Have we reached the threshold to say this is a causal relationship? For sun exposure and MS, I would say yes."

And not just MS. The effect shows up in related diseases, too. "There's a consistency of evidence across autoimmune diseases that have a similar immunopathology," Lucas says. "We've now shown it in pediatric MS. We've shown it in Crohn's disease. We've shown it in type 1 diabetes."

It also shows up in childhood eczema and allergies. Eczema is a form of chronic skin inflammation caused by the immune system's overreacting to common allergens in the environment. It's common in infants and has been on the rise since the 1970s. But when a team of Australian researchers attached light monitors to babies, they found that sun exposure dramatically reduced rates of eczema. (Vitamin D supplements, as usual, made no difference.)

Those sunnier babies also had lower levels of general inflammation in their bodies, which was also true for the MS patients in Prue Hart's clinical trial who received the UV light treatment. Inflammation is a defining characteristic of autoimmunity—it's a sign of the immune system on the warpath—but it's also the wormhole that we can follow to an entirely different perspective on why sunlight might boost health in so many nonobvious ways.

WE TEND TO THINK of inflammation as a bad thing. We treat it with ice and Advil after a workout, or we take more serious drugs and adopt anti-inflammatory diets to make it go away. But inflammation in itself is not a bad thing. Quite the contrary. When you're under attack from an infection, you need to call in the army. The heat and swelling of inflammation is a sign of the battle, as the immune system rushes defense forces to the site to destroy the invaders. Some friendly fire is unavoidable, but that's okay. Without inflammation, the kingdom would be lost every day.

Once the battle is won, the immune system is supposed to switch to anti-inflammatory mode. Turn off the flamethrowers, clean up the debris, heal the damage, retire the shock troops. (Immune responses are damaging and use a lot of resources. You don't want to live under martial law unless you have to.)

That elimination of dangerous pathogens is the canonical role of the immune system, but it's actually just a subset of the overall job description, which is to deal with any type of stressor that threatens to throw the body out of *homeostasis*, that Goldilocks zone where all systems are purring along in their happy place. That stress might come from a microbe or a cancer cell. But it could also be a regular cell that gets damaged in daily life. When the stress of exercise causes muscle cells to rupture, it's up to the immune system to clean up the mess and make the repairs. The heat and swelling of sore muscle after a workout is the same inflammation as that around a wound.

So inflammation is healthy and is the first stage of healing, and you can think of healing as the greater goal of the immune response. But sometimes something goes wrong. Maybe the bad guys keep invading. Or they go underground and never get entirely smoked out. Or our good cells keep going bad. Or maybe the immune system mistakes perfectly good cells for bad ones.

That can manifest as one of the autoimmune diseases, each of which involves the immune system attacking a different part of the body. In MS, it's the myelin sheaths. In type 1 diabetes, the pancreas. In eczema and psoriasis, it's the skin cells. In Crohn's, the lining of the intestine.

Rates of autoimmune diseases have been rising for years. Something about the modern world is throwing our immune systems out of whack. There's no shortage of culprits: Pollution, chemicals, sedentariness, and stressful lives may all play a part.

But surprisingly, one of the strongest factors influencing the behavior of the immune system is light. I can say this with confidence because it's been well documented for more than fifty years by researchers in some of the world's top medical institutions. The field even has its own name: *photoimmunology*.

Yet despite light's powerful impact on health, photoimmunology is almost completely unknown to the general public, or even to scientists outside the field. It's one of the sharpest examples of the science silos that I mentioned at the beginning of this book.

I'd certainly never heard of photoimmunology until I started showing up at light conferences. But once I did, it revolutionized my thinking. I began asking different questions. How does the body handle the daily barrage of photons? How does it use the information in those photons to get a sense of the outer world and to prepare for the future?

And what happens when the lightscape turns supremely weird?

IRONICALLY, THE TRAIL THAT led scientists to the discovery of UV's beneficial immune effects began from the confirmation of its dangers. In 1974, Margaret Kripke—the same immunologist who twenty years later would document sunscreen's inability to prevent melanoma in mice—discovered that tumors she induced in the skin of mice with a UV lamp failed to grow when transferred to the skin of a new mouse. The new host's immune system quickly eliminated them. Ten times she tried, and ten times the tumors were squelched. Only once she suppressed the new host's immune system with drugs did the tumor take hold. "That was the key!" she later recalled.

But why was the tumor able to grow in the original irradiated mouse? Was the UV light that had induced it also somehow suppressing the mouse's natural immune response? Indeed, in a series of experiments, Kripke determined that UV radiation was a double whammy. Not only did it damage DNA and trigger mutations that could lead to skin cancer, it also suppressed the immune system's surveillance of the skin, preventing it from killing any baby cancers that arose.*

This was a breakthrough in our understanding of how skin cancer develops, but from an evolutionary perspective it seemed nonsensical. "The fact that photons in this range have any effect at all on the immune system is both unexpected and remarkable," Kripke wrote. "It is unexpected because UV radiation has little power of penetration through living tissues, and most is absorbed by the outer few millimeters of the skin. It is remarkable because life has evolved in an environment containing UV radiation, and there is no obvious reason why the immune system should be influenced by it." Indeed, how could we ever have evolved an immune system that relaxed in response to a potential carcinogen?

Part of the answer is that back in our evolutionary past it probably wasn't a problem. As we've already seen, our dark-skinned forebears had no issue with skin cancer. The combination of thick skin, copious melanin, excellent damage-repair systems, and a protective layer of sun-adapted microbes would have prevented virtually every DNA mutation that might lead to cancer. In 2025, a team at the University of Lyon even discovered that urocanic acid (UCA), one of the main molecules that is produced in the skin by sunlight and is responsible for immunosuppression, is a favored food of sun-loving skin bacteria, which gobble it before it can trigger immunosuppression. Tumors

* Later studies found that broad-spectrum sunscreens prevented immunosuppression, but sunscreens that didn't block UVA failed to, bolstering the theory that those earlier generations of sunscreens contributed to the rise of melanoma.

flourished mostly in skin that lacked these bacteria—which would have been rare in the prehistoric world.

Other researchers have learned that the immunosuppression is directed mostly at the part of the system known as the *adaptive* immune system, while light seems to stimulate the other part of the system known as the *innate* immune system. The adaptive immune system is the sophisticated part that learns to recognize the unique genetic signature of particular threats, whether pathogens or cancerous cells, and destroys them. The innate immune system deals with immediate cellular damage of all kinds, regardless of cause, walling off wounds, eliminating compromised cells, and making repairs.

In that view, the immune system's whole response to sun exposure starts to make sense. In the natural world, exposure to mysterious molecules is constant. Most of these are harmless: garden-variety microbes, dust, pollen. Microscopic skin abrasions are also constant in a world of insects and scratchy surfaces.

In such an environment, the immune system has to learn not to overreact. What's a real threat, and what's an everyday occurrence? Exposures that don't lead to trouble eventually get ignored, but just like a new cop on the beat, a naive immune system gets better as it encounters more examples of what's trouble and what's not.

In the modern world, we see this with peanuts. Kids introduced to small amounts of peanuts at an early age are less likely to develop full-blown allergic reactions. Or farms: Kids exposed daily to garden-variety microbes are less likely to develop allergies and asthma compared to urban kids raised in more sterile environments. Those of us who live in blackfly country have noticed this dynamic with bug bites, too: The same bites that raise itchy welts on tenderfoots from the city barely register with leathery locals who have been nipped ten thousand times before.

For millions of years, solar radiation was one of those daily irritations, scuffing the skin and keeping the immune system engaged. "It's a challenge to the body," says Prue Hart, the Australian immunologist. "It's the most important environmental insult we have. We evolved to cope with it."

And to be tolerant. Instead of living with constant inflammation, beset with rashes and hives, our systems learned to hold their fire. It's a highly sophisticated and precise response. The big guns of the immune system get the message to stand down, while local repair systems go to work fixing the minor damage. Life goes on. The trade-off—especially today, with our sun-naive skin and suboptimal microbiomes—is that every now and then a new cancer sneaks through the defenses.*

THE DISCOVERY OF UV'S powerful impact on immunity launched the discipline of photoimmunology. "Before our studies on the immunology of skin cancer, no one would have dreamed that shining a light on the skin would have immunological consequences," Kripke reflected years later, "much less that this would turn into an entire new field of investigation."

Understandably, photoimmunologists originally focused on the negative effects of UV immunosuppression. But they soon found upsides as well. For example, it finally explained something people had noticed for centuries: Sunlight soothed psoriasis, a skin condition marked by painful, itchy scales. Back then doctors didn't understand why light improved the symptoms, but we now know that

* One fascinating bit of confirmation of this idea comes from people who suffer from polymorphic light eruption (PLE), a common disorder in which patients' immune systems *do* respond forcefully to sunlight. PLE sufferers get itchy rashes and plaques after sun exposure, but they are less likely to develop skin cancer.

psoriasis is caused by an ongoing immune attack on the body's skin cells. Ultraviolet light—whether from the sun or a lamp—improves things by tamping down the immune response.

Although some doctors do tell their psoriasis patients to simply get more sun exposure, more often they recommend narrowband UV therapy instead. UV does most of its direct damage to DNA at wavelengths around 290–300 nanometers, but a UV lamp with a wavelength focused around 311 nanometers seems to deliver the benefits without raising the risk of skin cancer. The same treatment is used for vitiligo, an autoimmune disease of the skin caused by the immune system attacking the melanin-producing melanocytes, resulting in blotchy white patches of depigmented skin.

The use of UV lights to treat psoriasis and vitiligo in millions of patients eventually made it obvious that the effects of the light went more than skin deep. "You get this rebalancing," says Prue Hart. "UV calms inflammation in the skin. But it also then calms inflammation in the central nervous system. It'll calm inflammation in the pancreas and the gut. So I think it's not fully realized the potential it has to be a controller of body homeostasis."

Some of the best research comes out of Korea. A 2021 study of nearly a million psoriasis patients found that those who used phototherapy had 36 percent fewer major cardiovascular events, including heart attacks and strokes, than those who didn't use the therapy. That unexpected finding was confirmed by another 2021 study of more than twelve thousand vitiligo patients, which found that those who had received regular phototherapy treatments for years were 32 percent less likely to suffer heart attacks and 40 percent less likely to suffer strokes compared to those who didn't use phototherapy.

That helped confirm not just what the photoimmunologists were seeing but also the cardioprotective effects of sun exposure documented by Richard Weller and Pelle Lindqvist. But it also begs a new question: Why would turning the dials on the immune system influence *cardio* health? That's not an autoimmune condition.

Or is it? In recent years, many seemingly unrelated diseases have turned out to have inflammatory roots. Cardiovascular disease is often caused by immune cells attacking and damaging the walls of the blood vessels. Alzheimer's is connected to low-grade, smoldering inflammation in the brain. Many COVID deaths were caused by massive inflammation in the lungs.* Arthritis, asthma, diabetes, cancer, even depression, are thought to have inflammatory components.

The underlying issue is that our bodies now exist in chronic low-grade agitation. The problem is so ubiquitous that the journal *Cells* recently ran a special issue titled "Inflammation: The Cause of All Diseases."

And the problem gets worse as we age. This *inflammaging*, as it's been termed—a kind of smoldering immunological fire that won't go out—shows up in all of us in the modern world. In the broadest sense, inflammation is simply the overall efforts of the body to return tissues to their normal state. Inflammaging is a sign that, for whatever reason, lots of our cells are a little bit off—or at least the immune system treats them that way.

But interestingly, inflammaging isn't seen in hunter-gatherers of any age. Part of the reason seems to be that their immune systems are already fully occupied dealing with the world. Hunter-gatherers are exposed to a lot more pathogens and parasites through a wild diet and a life lived in the bush. Like a talented but mischievous teenager, the immune system needs to be kept busy with useful things to do or it will find things to do, useful or not.

Hunter-gatherers are also exposed more than most city dwellers to two other forms of daily stress that engage their immune response:

* And they were less frequent in sunnier latitudes.

exercise and sunlight. I mention them together for a reason. The list of conditions the two factors address is nearly the same: heart disease, stroke, diabetes, high blood pressure, atherosclerosis, cancers, depression, anxiety, dementia, obesity, osteoporosis, osteoarthritis, sleep disorders, and autoimmune diseases. Maybe that's not accidental. Maybe they are two versions of the same fundamental process.

EXERCISE IS FUNNY. We've known for so long that it's good for us that we rarely stop to consider why. When we do, suddenly it's not so self-evident. In the moment of doing it, exercise isn't at all good for you. This is obvious anytime you push your limits. It's hugely stressful. Your lungs heave. Your heart pounds. Ligaments stretch. Joints take a beating. Your convulsing muscle cells burn oxygen at a hundred times their resting rate. Molecules are being combined and ripped apart at unfathomable speeds. Damage is unavoidable. Proteins shred. Electrons fly in all directions. Cells explode.

How could any of that be good for you?

The answer is that it all gets fixed. The body's repair mechanisms swing into action. That starts with inflammation, as the immune system rushes its forces and supplies to an area to fix the damage. The swelling and "burn" after exercise is the immune system in damage-control mode.* That's followed by an anti-inflammatory phase, in which the immune system puts out the fire. And while the firefighters are deployed, they put out any other fires they find. In addition, signals go out to the body to build more capacity for next time: more muscle, more oxygen transport, more energy-generating mitochondria, more damage control.

So, counterintuitively, exercise is anti-inflammatory in the long run because it's pro-inflammatory in the short run. It's a fire drill, a stress

* And it should not be arrested with ice or NSAIDs, FYI. It needs to run its course.

test, and its effects on health are so broad because it prompts the body to put out smoldering fires that have nothing to do with the exercise.

But it has to be exercise with the right amount of challenge. Walking from the couch to the fridge isn't going to build capacity, but neither is running until your arches collapse.

It may seem odd that we have such a screwy system in place. Why not just have the damage-repair crew kick in every day automatically, without the need for a stress trigger? The answer is probably that historically we had no option but to exercise for hours every day, and our immune system evolved in that context. Only with the numbing understimulation of modern indoor existence could this ever be a problem.

And though exercise is the gold standard of good stress, it's far from the only example of *hormesis*, as the wellness gurus call it. We also get better at snuffing out inflammaging daily through exposure to garden-variety microbes, the phytochemicals in plant foods,* the temperature extremes of saunas and cold plunges, intermittent fasting—and sunlight. Sun exposure improves the same health conditions as exercise because it *is* a form of exercise. It's a workout for the skin, best in moderation as part of an active lifestyle of good food, good movement, and good light.

UP TO NOW, I've focused almost exclusively on interactions with ultraviolet light, with good reason. UV is the only wavelength with enough energy to vaporize atomic bonds. That makes it high risk

* Part of the reason plant-rich diets such as the Mediterranean diet are so good for us is because they are full of mildly toxic compounds designed to discourage animals from munching on them too much. But as Paracelsus observed five hundred years ago, the dose makes the poison, and a soupçon of those tingly toxins serves as mild, medicinal stress.

and high reward. It can produce vitamin D and nitric oxide and other beneficial molecules in the skin. It can also zap DNA and charge up free radicals, both of which can lead to skin cancer. So it gets the lion's share of the attention, and biology goes to great lengths to deal with it.

But as we know, sunlight is much more than UV, which comprises just 4 percent of solar radiation. Visible light covers 43 percent, and infrared the other 53 percent. All these different wavelengths have different energy levels, different abilities to penetrate the skin, different affinities for different molecules, and unique effects on mind and physiology. Visible light and infrared have been mostly ignored until recently, as if they were nothing but stage lighting. But to understand our full relationship with light, you have to look at the whole picture.

That picture is filling in fast, making it abundantly clear how vital these energy forms are to our daily health, and how risky it is to drastically alter the lightscapes of our environment without any consideration of the consequences. Through the eyes, the visible wavelengths have major impacts on mood, consciousness, circadian rhythms, sleep, and performance. They're so important to daily functionality that they'll be the focus of the final part of this book.

Before we get there, however, we need to turn to the most abundant and yet the most overlooked wavelength. Infrared light gets ignored because it's invisible and because it has the lowest energy of the solar wavelengths. Not only can't it break a molecular bond, it can barely excite a molecule at all. It just seems to slip inside us, lurking everywhere, a background hum. But that gentleness may be the secret to its power.

CHAPTER 10

IN THE RED ZONE

The Pervasive and Perplexing Powers of Infrared Light

One of my favorite feelings in the world is to stretch out in the sun on a late-winter day and to feel its heat slowly warming me deep in my bones. Until recently I'd have said that the "deep in my bones" part was metaphorical, but amazingly, scientists are now learning that some of those solar photons do indeed penetrate all the way to my bones, as well as my other cells, making them not just warmer but healthier, too. And though the science is young, it's inspiring a reconsideration of why natural sunlight might be good for us—and why a deficit might result in "modern-day scurvy," as one researcher puts it.

If you haven't read it already, now would be a great time to check out the color insert "A Light Primer," which gives an in-depth explanation of the physics of light. But here's a quick refresher: When a photon of light meets a molecule of matter, it can either bounce off, pass through, or be absorbed. If the photon and the molecule are oscillating at a similar frequency, the photon and its energy will be absorbed by the molecule. If they are worlds apart, the photon will pass by as if the molecule didn't exist. And if they are close but not too close, the photon will bounce away in a new direction.

It's easy to picture this with a pane of glass. Photons in the visible range will pass right through. Add a mirror coating to the glass, and they bounce back. Paint it black, and they are absorbed.

This happens in living bodies, too. Depending on how a particular photon and molecule match up, the light might bounce away, be absorbed, or glance off and continue its billiard-ball path farther into the body.

As a general rule, the shorter the wavelength of light, the less distance it travels into the body. Ultraviolet light, which has the shortest wavelengths of the solar spectrum, never makes it beyond the outermost layer of skin before meeting a molecule that absorbs it. Blue and violet, the shortest of the visible wavelengths, make it a little deeper before being absorbed. Red light, the longest of the visible wavelengths, penetrates a bit beyond the skin.

But just beyond red lie the gentle infrared wavelengths known as *near infrared*. (Red light includes wavelengths of 620–780 nanometers, which is the limit of what our eyes can detect. Near infrared covers the next stretch, from 780 to 1,400 nanometers. Mid-infrared goes from 1,400 to 3,000 nanometers, and far infrared—the kind used in infrared saunas—goes all the way from 3,000 to 100,000 nanometers.) Few molecules in the body oscillate at the right frequency to absorb near-infrared photons, which travel deeply into the body, ricocheting this way and that until they either exit or meet a molecule that can absorb them.

The main molecules in the body that successfully absorb near-infrared light are found in mitochondria, the tiny bacteria-like structures in cells that handle many of the basic tasks of life, including energy production. That puts them upstream for pretty much every other function of the body, from motion and cognition to healing. If they wear out, so do we. Today, many of the classic diseases associated with aging—cancer, cardiovascular disease, chronic inflammation, maybe aging itself—are being traced to mitochondrial dysfunction.

How do we keep our mitochondria healthy? Well, all the usual suspects—a good diet, regular exercise, avoiding toxins—but red and near-infrared light may be another key. That field of research is officially known as *photobiomodulation*, though it's often just called red-light therapy, and it's now used every day in thousands of clinics around the world to treat everything from wound healing and hair loss to cognitive disorders such as Parkinson's disease.

But as evidence for its efficacy grows, it also begs some basic questions about the role of infrared light in everyday health. As mentioned, infrared light comprises 53 percent of the solar spectrum—more than all the visible light combined. Anytime we are outside in daylight, those photons are pouring into us, delivering their energy to our mitochondria, even right through clothing. That used to happen all day, every day. Now it happens only when we are outside. And a growing number of researchers are becoming deeply concerned about this drastic shift in our energy diet.

FOR A LONG TIME, I kept photobiomodulation at arm's length. Sure, it has enthusiasts everywhere from NASA to the American Academy of Dermatology, but for me there were too many classic signs of quackery. Strange gizmos. High-tech explanations that didn't quite make sense. Literal flashing red lights.

Typical devotees can be seen strapping on helmets fitted with red LEDs *on the inside*, the better to inject the healing rays into their brains, or taping what appear to be red laser pointers into their nostrils, causing their sinuses to glow like lava lamps. Biggest quackery flag of all: sketchy online companies eager to sell you devices.

But when you push aside the hucksters and just look at the work being done in clinics and universities around the world, you find both a decent track record of success and some solid, if still unproven, explanations for why it works.

The field traces its origins to a Hungarian surgeon at Semmelweis University in Budapest named Endre Mester. In 1965 Mester was experimenting to see if a laser could be used to destroy tumors. A laser is a highly focused light, capable of delivering a lot of energy to a precise target. The very first one had been invented just five years earlier, and a guy in Boston had reported success treating tumors a couple of years later, so Mester decided to give it a go. He stitched tumors beneath the skin of rats and zapped half of them with a ruby laser he'd made himself. To his disappointment, the laser had no effect on the tumors. But unexpectedly, the incisions healed faster in the rats hit with the laser. A lot faster.

Mester was intrigued, so he investigated further, discovering that his laser accelerated the healing of all kinds of wounds, scars, burns, and ulcers. It seemed to work best on distressed tissue. Shave the back of a mouse, apply a daily dose of laser light, and the hair regrew faster. Mester was stumped. Why would a searing blast of energy heal, rather than vaporize?

As it turned out, the fault lay in Mester's homemade laser, which was much less powerful than he thought. He'd given those rats more of a kiss than a zap. And like most kisses, it was highly therapeutic. Many other experimenters as well found that it worked, and that you didn't even need to use a laser—simple red or near-infrared light worked, too.

No one had a clue why it might work until the 1980s, when an Estonian scientist named Tiina Karu discovered that certain molecules found in the energy-producing units of mitochondria had just the right structure to absorb deep-red and near-infrared photons.

Karu proposed that this extra boost allowed mitochondria to produce energy more efficiently, which led to less wear and tear on them and more longevity for the cell. (When mitochondria fail, cells die.)

The whole field might have remained fringe had it not gained the backing of a decidedly non-fringe player in the 1990s. NASA had a long-term interest in finding ways for astronauts to grow their own food in space, and in 1995 it began trials to grow potato plants on the Space Shuttle. Using typical full-spectrum grow lights on the plants was not an option, as the lights are too hot and energy intensive. So NASA came up with an energy-efficient LED system using just red and blue—the two wavelengths plants absorb for photosynthesis.

One of the many challenges of space flight is that astronauts' wounds heal more slowly than on Earth. A number of culprits have been proposed—maybe it's the extra radiation astronauts are exposed to in space, maybe skin cells and immune cells can't coordinate as well in microgravity—but if you have made it this far in this book, you will suspect that lack of natural light might be a factor, and you'll be right.

To everyone's surprise, abrasions on the hands of astronauts who tended the plants in the red light healed significantly faster. NASA pivoted and made red-light therapy the focus of the project, spending the next eight years documenting its effectiveness.

That helped the technology gain credence in the mainstream medical world. Most of the enthusiasm has come in fields where the red light can directly touch tissue that needs healing. My dental hygienist used a red laser on me in 2025 without fanfare, though when I chirped excitedly and peppered her with questions, she had no idea why it was supposed to work. She just knew that gums healed faster after a deep cleaning when kissed with the red light.

One of the biggest surprises to me is that the American Academy of Dermatology is all in on photobiomodulation. In 2024, Henry Lim, the former president of the academy, told a roomful of dermatologists at the academy's annual meeting, "It is essential for us to continue to be the owner of this subspecialty." Apparently he was worried about losing it to the guys sticking laser pointers up their noses. Lim coauthored a pair of articles in the academy's journal explaining the science and documenting its ability to attenuate pain and inflammation, rejuvenate hair growth, and accelerate wound healing.*

But derms do have to share the subspecialty with many others. It's being used to reduce pain and inflammation in deep tissue and joints, possibly because infrared light has shown some ability to release nitric oxide in cells. Infrared light also seems to improve the function of lung cells, and it's now being researched as a treatment for respiratory diseases including COVID, pneumonia, and respiratory distress syndrome. Curiously, it has also shown promise in a variety of neurological disorders including Alzheimer's, Parkinson's, stroke, depression, and traumatic brain injury, where the noninvasiveness of the treatment—just shine a light into people's skulls—would be most welcome.

But that also leaves a tantalizing mystery. It's one thing to use red or infrared light to treat skin or gums, or even injuries slightly beneath the surface of the skin. But to affect the brain, light would have to penetrate a lot more than skin. Until recently, no one thought it did.

ONE OF THE PEOPLE who got curious about photobiomodulation was Scott Zimmerman, a New Jersey optics engineer and inventor with dozens of patents to his name. Zimmerman is a self-described nerd, with no particular interest in alternative medicine, but optics

* Along with nitric oxide generation, this may help to explain the observations by surgeons in World War I that wounds healed faster with sunlight.

A LIGHT PRIMER Some Notes on the Physics of Photons

Light is pure energy, traveling through the universe at 300 million meters per second. It has no mass at all. It comes bundled in discrete packets called photons. Virtually all the light in the universe is produced by stars, including the sun, which release huge amounts of energy in the form of photons during the intense nuclear fusion of hydrogen atoms that takes place in their cores.

Light travels in waves—pulses of energy that flow through space the way an ocean wave flows through water molecules. Depending on how much energy it has, a light wave will oscillate at a particular wavelength as it travels. The more energy, the shorter the wavelength.

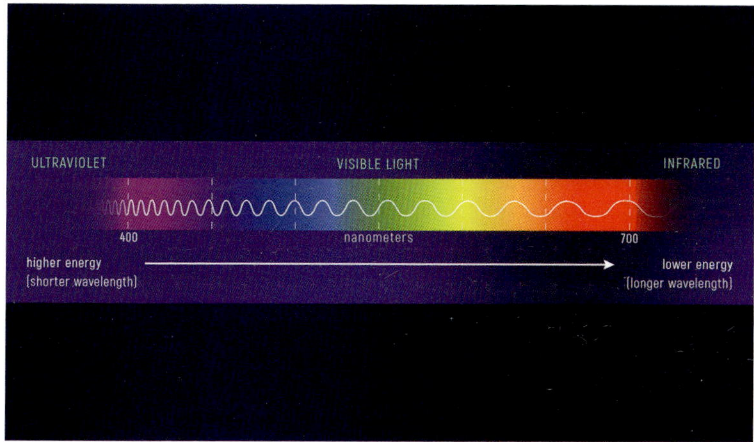

NASA

The light that reaches the Earth's surface comes in a spectrum of wavelengths from 280 nanometers (nm) to 2,500 nm. Our eyes can detect light with wavelengths between 380 and 780 nm. We call that "visible light," and we perceive the combination of all those wavelengths as "white," but it actually contains all the colors, which becomes obvious when you shine sunlight through a prism. (Or when water in the atmosphere acts like a natural prism to produce a rainbow.)

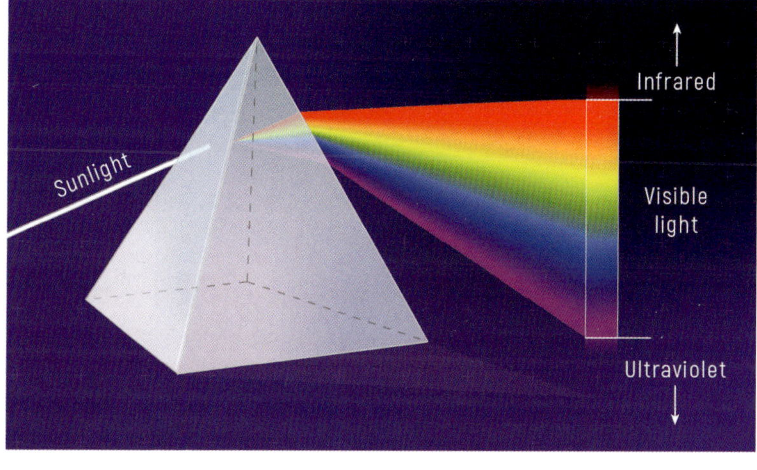

NASA

Visible light makes up just 49 percent of the sunlight that reaches us. Of the rest, 4 percent is ultraviolet ("beyond violet"), which includes the high-energy light with wavelengths between 280 and 400 nm, and 53 percent is infrared ("below red"), which includes the low-energy light with wavelengths longer than 750 nm. We don't have receptors in our eyes for these wavelengths, so we can't see them, but our cells can still detect them and absorb their energy, and they have profound biological effects. What we perceive as different colors are just light rays with different wavelengths and energy levels. Those colors don't exist in the physical world; they are mental images we build in our brain using the electrical signals sent by the receptors in our eyes. We have three different models of receptors in our eyes, each capable of absorbing a different range of wavelengths.

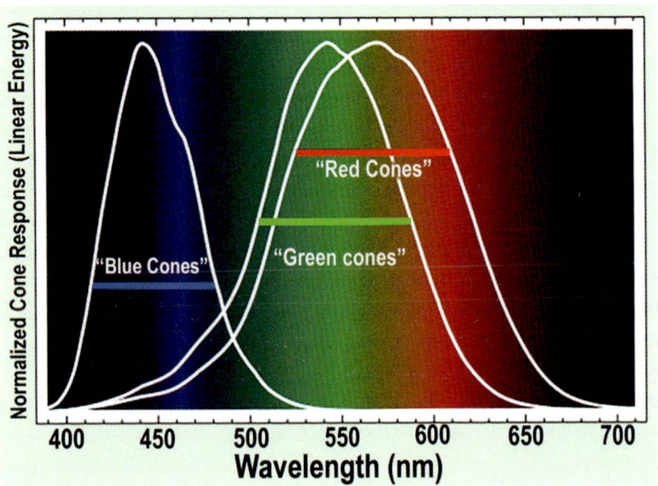

NASA

One of these three receptors is tuned to shorter visual wavelengths (blue), one is centered on medium wavelengths (green), and one is best at long wavelengths (red). All the colors we see are made in the brain by combining the electrical signals coming in from these receptors. If they are more or less balanced, the brain reads that as "white" light. But fundamentally, all that makes one color different from another is energy and wavelength.

When we see an object, what we are actually seeing is the photons of light that are bouncing off that object and entering our eyes. In other words, what we see is the stuff that wasn't absorbed. So an apple is red because its skin is reflecting red photons (those with wavelengths around 650 nm) while absorbing the other colors. As you navigate the world, whatever colors you see tell you which wavelengths of photons are hitting the rest of you, too. (In addition to ultraviolet and infrared, which you can't see but might be hitting you.)

When a photon of light does encounter a molecule of matter, it can pass through, bounce off, deflect, or be absorbed and disappear, transferring its energy to the molecule.

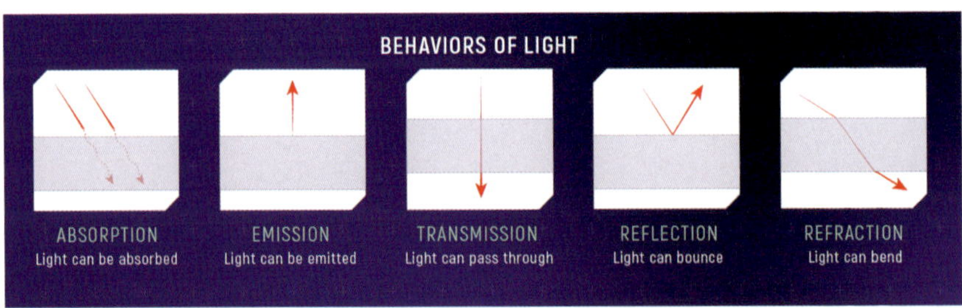

NASA

Which of these happens depends on the wavelength of the photon and the oscillation of the molecule. In the atmosphere, shorter wavelengths are more easily deflected, or "scattered," than longer wavelengths. The sky looks blue because those are the photons most easily bouncing around the atmosphere before reaching us. The more atmosphere the light must pass through, the more photons get scattered. We see this every evening at sunset, when red and orange light comes straight at us but the other colors are bent away.

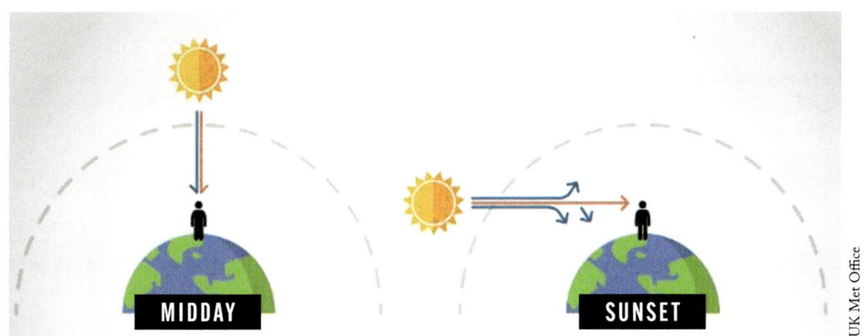

UK Met Office

Molecules also oscillate at a certain frequency, depending on how much energy they contain. If a photon hits a molecule with a complementary amount of energy, the photon will be absorbed. When photons of light hit our skin, some bounce off, while others penetrate until they hit a molecule that can absorb them. The shorter the wavelength of a photon, the more likely it is to quickly run into a molecule. Ultraviolet light with wavelengths of 280 to 315 nm ("UVB") doesn't even make it through the epidermis, the thin outermost layer. UVA (315 to 400 nm) penetrates a little bit farther, blue farther still, and so on. "Near infrared" penetrates the farthest of all.

In fact, because there are very few molecules in the body that absorb near-infrared photons with wavelengths around 800 to 900 nm, some of these photons can travel through the entire body. Here is a photograph by the astrophysicist Bob Fosbury of an infrared light shining through his hand (and photographed with an infrared camera). The light comes straight through the flesh and bones, which can't be seen.

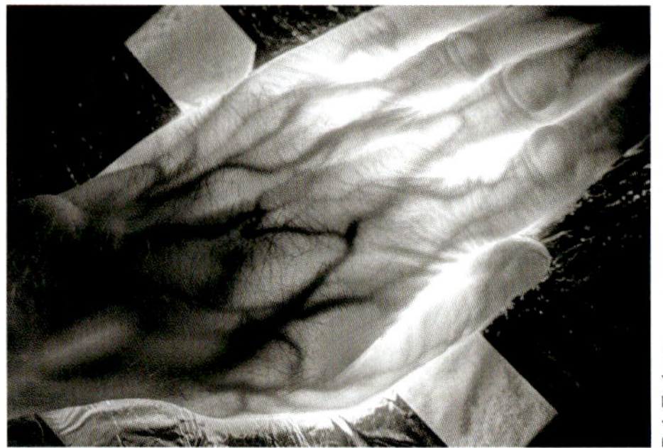

Bob Fosbury

Different wavelengths of light have different effects on the body. Ultraviolet photons have so much energy that they can actually break the atomic bonds that hold molecules together. This can be used to produce beneficial new molecules like nitric oxide and vitamin D, but it can also damage DNA and lead to skin cancer. Visible photons don't have enough energy to break bonds, but they can make molecules twitch. That's what happens with the receptors in our eyes. When they absorb the energy of the right photon, they twitch and send an electrical signal to the brain. Infrared photons have very low energy. They can't make a molecule jump, but they can make it a little bit more active. Bob Fosbury compares this to "ringing a bell" or "making them sing."

Where we live has a huge impact on how much light we are exposed to. Because of the tilt and curvature of the Earth, the sun appears lower in the sky in high-latitude locations, and its light must pass through more atmosphere to reach the surface. The more atmosphere, the more photons get filtered out. In addition, the amount of cloud cover and moisture in the air affects how much light reaches the surface. The result is a lot of variation in how much light energy reaches different locations. While the American Southwest or Australia might receive 2,500 kWh of solar energy per square meter each year, New York might get 1,500; Paris 1,000; and Scotland just 600. This is reflected in different rates of disease for many conditions affected by sunlight.

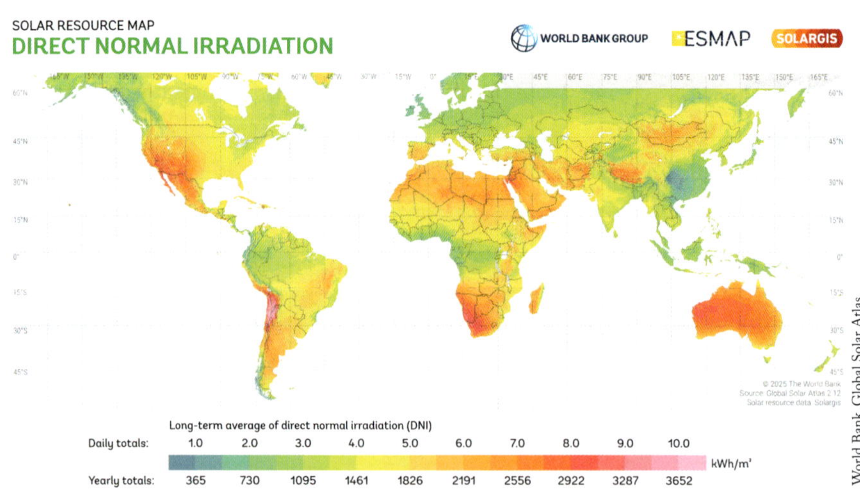

World Bank, Global Solar Atlas

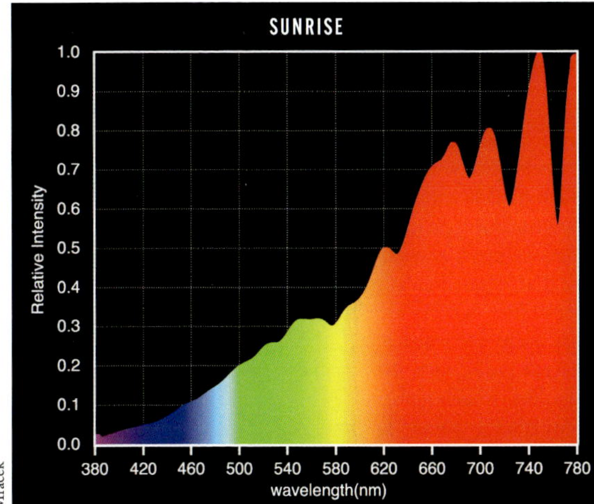

Mracek

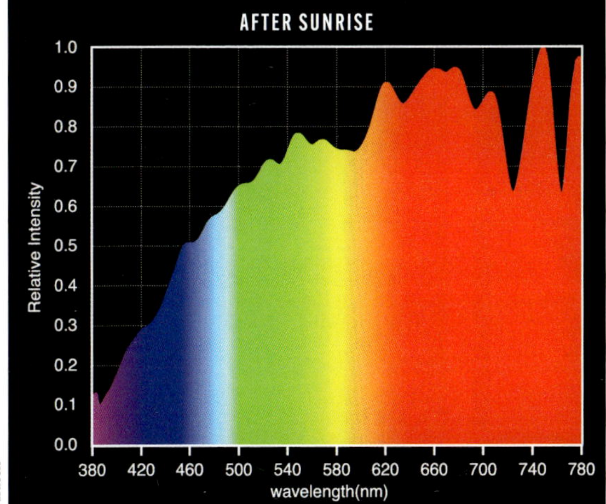

Mracek

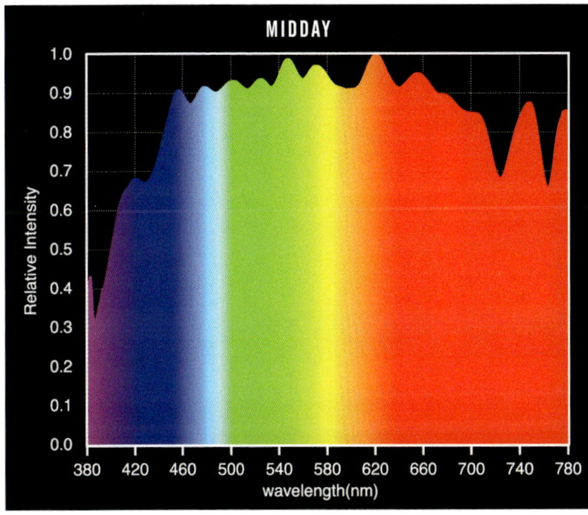

Mracek

But in addition to location, other factors such as season, time of day, and type of lighting can have a big impact in the amount and kinds of photons we are exposed to. For example, on this page are spectrograms showing the mix of light at three different times of day: sunrise (or sunset); after sunrise (or before sunset), when the sun is up but still low in the sky; and midday. The different wavelengths (colors) are shown along the X axis and the total amount of each wavelength is shown along the Y axis.

At sunrise, most of the light is in the red zone, because the shorter wavelengths are being filtered out by the atmosphere. Although not shown in these spectrograms, ultraviolet light in particular (below 400 nm) is almost entirely eliminated. This is why morning and evening light doesn't cause skin damage and can be particularly beneficial, bathing the body in gentle, longer-wavelength energy. In contrast, the midday spectrum of natural daylight includes a broad distribution of all the visible wavelengths, as well as ultraviolet and infrared.

You can see that natural daylight has a lot of blue in it, and blue is the cue for our brains to stop making melatonin (the hormone that makes us sleepy) and to become highly alert for the day. As the sun begins to set, the blue disappears, and soon melatonin production begins again. In prehistoric times, when we had no man-made sources of light, that was that.

Now look at the spectra from four other sources of light: fire, an incandescent bulb, a compact fluorescent, and a high-quality LED designed to approximate natural daylight.

You can see that fire produces almost no blue light at all, so it doesn't suppress melatonin production. On the contrary, as we all know, fires are pretty good at making us sleepy, probably a combination of their heat and the warm bath of all that infrared radiation. The old incandescent lightbulbs also produced a spectrum of light very close to fire—a lucky break for us, because once again they caused little melatonin suppression. They were warm and comforting.

Then along came compact fluorescent bulbs, with their freakish, spiky spectrum. God knows what they did to our brains. Thankfully, they've been replaced by LEDs. But LEDs are very different from incandescents or fire. The ones designed to mimic "bright daylight" contain a lot of blue light, as you can see in this high-quality Mracek 5000K bulb. For an artificial lighting source, it does a remarkably good job of approximating the solar spectrum, which makes it an excellent choice for daytime activities where you want to be fully alert. At night, however, all that blue is going to suppress melatonin and interfere with sleep. A better bet for nighttime use are "warm white" LEDs (2,700K or so), which produce less blue and more green and red.

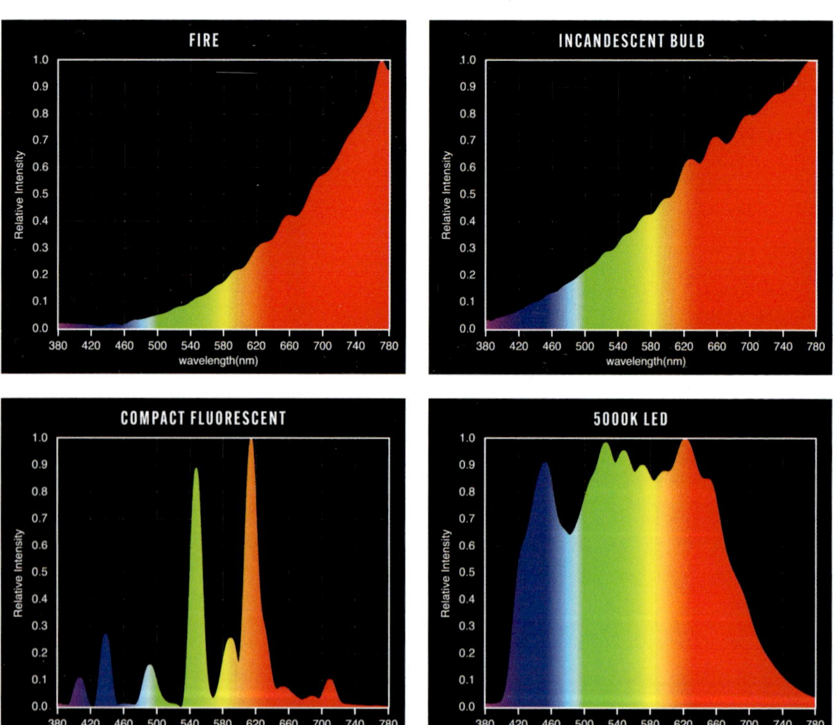

Mracek

Another important factor in light quality is shade and vegetation. Plants absorb red, blue, and ultraviolet light, but they reflect green and infrared. Here is a photograph by Bob Fosbury of a park in summer, taken with an infrared camera. You can see that the ground and bridge absorb infrared light but the plants reflect it. (Buildings also absorb infrared.)

Bob Fosbury

This means that when you are in the shade of a forest, you are protected from ultraviolet light but are receiving lots of other beneficial photons, especially those in the green and infrared range. Since those wavelengths in particular have been shown to reduce pain, stress, and anxiety, this helps explain why forested environments feel so good to us.

The upshot of all this light physics: Different wavelengths do different things to the body. As with nutrients, we need at least a little bit of all of them—and we can maximize our health by taking control of our personal light diet. Here's a quick reference guide to the suspected effects.

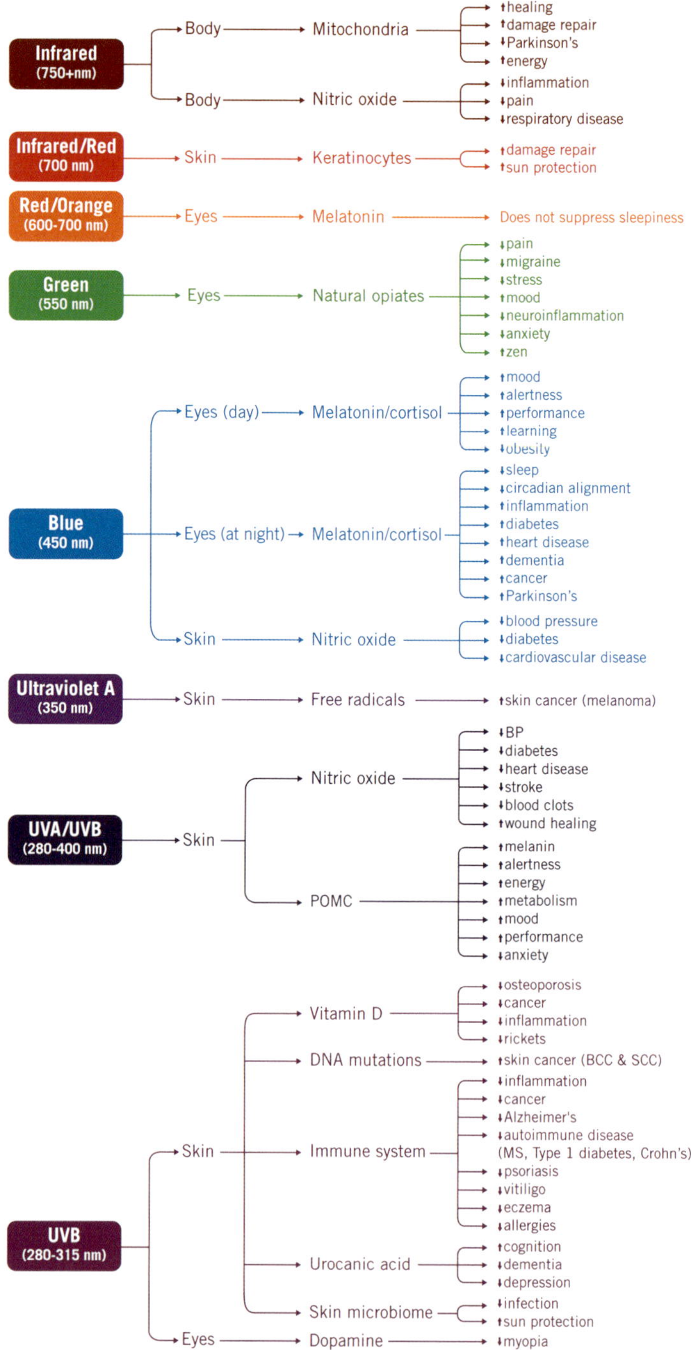

is his thing, and he was intrigued that photobiomodulation was being used to treat conditions that were more than skin-deep. Was infrared light better at penetrating living tissue than we knew? And did it matter?

To find out, he did what an engineer does, visiting the experts, asking questions, trying to wrap his head around the mechanics. What quickly became clear to him was that most people involved had no clue what they were doing.

For starters, the amount of light they were delivering in clinics was unimpressive. "I kind of got in trouble," he told me, "because I said, 'Let me get this straight: You're basically just accelerating a process that's going to happen naturally if we go outside?'"

Indeed, the amount of infrared light being wielded in many photobiomodulation experiments was just a slightly concentrated version of what anyone could get from good old-fashioned sunlight. That people were actually seeing benefits in the clinic suggested that they might not be getting enough in their daily lives—which was certainly possible since modern indoor lighting doesn't produce any infrared light at all.

Zimmerman was also perplexed by the use of photobiomodulation to treat deep-tissue injuries and even neurological symptoms. For that to happen, infrared light had to pass fairly easily through flesh and bone. Could that be the case?

He knew his first step to finding out. "I'm an optics guy. I needed to model the optics of the body. So I started looking around. I thought surely everybody's done this. And what I found was that nobody had."

No one with expertise in the physics of how light moves through a medium had done the right experiments to see how well various wavelengths could penetrate living tissue. Most biologists assumed no wavelength of light would make it much deeper than the skin.

One clue that it might was melanin, the tanning pigment that protects us from ultraviolet radiation. In addition to UV, the visible wavelengths are readily absorbed by melanin, as you can tell from its brownish-black color. But infrared gets a free pass. "Melanin has a peak absorption in the UV range," says Zimmerman, "but then it drops like a rock. It is so optimized that it almost disappears in the infrared range. And that is unique to its structure." To Zimmerman, it looked as if we had evolved to admit infrared.

Next he looked at *scatter*—the way light caroms through a medium. Hold a glass of red wine up to a light and you won't see the light coming straight through; it fractures in multiple directions. Zimmerman found that the same thing happens with near-infrared light and flesh. Once inside the body, the photons scatter like pachinko balls. Bouncing randomly this way and that, some of them find their way deep inside.

But flesh is one thing; bone is another. To affect the brain, infrared light would have to pass through bone.

Zimmerman got a hint that it could when he discovered the work of Bob Fosbury, a bespectacled astrophysicist in England who specializes in spectroscopy. Fosbury worked on the Hubble Telescope and built instruments for the James Webb scope. He spent his career analyzing the spectra of distant celestial bodies. In retirement, he turned his spectroscope to Earth. He began shooting photographs with an infrared camera and posting surreal images of everyday sights.

In one of his images, Fosbury shined an infrared light at his hand and took a photo from the other side with his infrared lens. The images showed light pouring through the hand. His finger bones were invisible, meaning the light had passed through them. The only things that were dark (indicating they had absorbed the infrared photons) were his veins, which looked like black rivers.

In 2025 Fosbury teamed up with Glen Jeffery, a neuroscientist at University College London, to test infrared's ability to pass through a body. They shined light at the chests of volunteers and checked for light emerging on the other side, using extremely sensitive detection equipment. Infrared got through every time. It wasn't much—just three or four photons out of every one hundred thousand—but it proved that near-infrared could penetrate the thickest part of the human body.

And for every photon that made it all the way through, many more were being absorbed deep inside. Fosbury estimates that somewhere between 10 and 50 percent of near-infrared photons that enter the body are absorbed before escaping. Considering the trillions of photons of sunlight that hit our bodies every second, that's a lot of shards of sun rattling around. We are, Fosbury says, walking light traps, shimmering with photons that seem to give our mitochondria more vigor.

Zimmerman has also found other signs that the human body might be optimized to make use of infrared light. "The more I dug," he says, "the more I could see all the things the body was doing to assist the light. It goes to great lengths to gather it."

The fluids in the eye, the womb, and the skull are exceptionally well suited to transmitting near-infrared photons. In the brain, Zimmerman has written, cerebrospinal fluid "acts as a light guide, distributing the near-infrared photons scattering off the surface of the brain even deep into the folds."

What might those cells be doing with this steady infusion of photons? Fosbury and Zimmerman are now investigating tiny nanostructures in brain cells that may hold the answer. In their design and layout, the nanostructures resemble photosynthesis molecules as well as solar panels. "It's exactly the same kind of structure and spacing that we put on solar cells to enhance their absorption characteristics," says Zimmerman. "Optically, it's one of the most beautiful things I've ever seen in my life."

The size, spacing, and absorption characteristics of these nanostructures look optimized to capture infrared photons. "We think these regular nanostructures are designed by evolution to trap that light," says Fosbury. The nanostructures are found on a part of the cell that provides nutrients to mitochondria. "It's like a service station for mitochondria," he says. "They visit it every day. They cycle around and get serviced. It's the source of raw materials so they can regenerate and repair themselves."

Fosbury believes one of the raw materials being supplied to them is light energy. The jolt of an infrared photon may be gentle, but it still carries about forty times the kinetic energy of a typical molecule in the body, which might be just enough to lubricate various metabolic processes, allowing mitochondria to do more with less and staving off their inevitable decline.

Infrared light may also improve the efficiency of mitochondria when it is absorbed by the water molecules in their energy-production centers, changing their vibrational frequency and allowing the energy to move through the system with less resistance.

"I call it making the molecules sing," Fosbury says. "There's an instantaneous effect. As soon as you go out in sunlight in the morning, suddenly your mitochondria work faster and generate more energy." Intriguingly, the wavelengths of infrared light that aren't absorbed in the atmosphere—that make it into our bodies—are also the photons that carry the exact amounts of energy required to give little boosts to the components of the mitochondrial energy system and help them along their way. To Fosbury, this is a clear indication that we evolved to take advantage of these little lagniappes of free energy.

The effect is enhanced in areas with lots of natural vegetation. In Fosbury's outdoor infrared photographs, buildings and other parts of the built environment look dark because they absorb infrared. But plant leaves shine brightly. Plants can't use infrared for photosynthesis, so they reflect it, and that means anytime we are in a green environment, we are getting an extra dose of infrared—once from the sky and a second bounce from the plants.

In contrast, indoors we get virtually none at all, even if the lights are on. Modern LEDs don't produce any infrared light. They are mostly blue, with small amounts of green and yellow—a mix that our brains read as "white." Starved of infrared, our mitochondria aren't singing at all. Fosbury calls this condition "modern-day scurvy" and believes it's at the root of many ills. "The insides of buildings are so toxic," he says.

BY THE TIME I'D gotten up to speed on infrared light, I was ready to agree with him. It was 2025, I'd made my way through hundreds of science papers and dozens of interviews, and I had little doubt that our disconnection from sunlight—that full-spectrum mix of ultraviolet, visible, and infrared—was lurking in many modern ills. I was also growing increasingly concerned about the other half of the twenty-four-hour cycle. Too much artificial light at night seemed to be as bad as too little sunlight during the day.

And I honestly wasn't doing particularly well on either front. No matter how much you know about the problem, a life of reading and writing and researching is still an indoor life. And I could tell it was all catching up with me. Until then, my curiosity about light and health had been a purely intellectual exercise. But it was about to get personal.

PART III

ALL THE LIGHT WE CAN'T UNSEE

CHAPTER 11

TICK TOCK

The Scourge of Artificial Light at Night

If you want a perch from which to observe the circadian train wreck of modern industrial society, you could do worse than a corner booth in the Tick Tock Diner. A 24/7 neon beacon holding down the corner of Thirty-Fourth Street and Eighth Avenue, the Tick Tock serves up Manhattan on a plate. Breakfast all day. Corned beef on rye. An actual Manhattan if you'd like. The rainbow columns of Madison Square Garden glow across the street. The train to everywhere is directly below. Down Thirty-Fourth, the Empire State Building fills the sky, its sixty-eight thousand LED lights pumping power into the cosmos. That pale smudge to the left? That's the moon.

When a Knicks game lets out at the Garden or concertgoers from the Meadowlands pour out of Penn Station, crowds pack the Tick Tock's orange-and-green banquettes and the line snakes out the door. But even in the thin hours, people find it. The cop warming her hands around a coffee. The Times Square revelers not ready to call it quits.

And me. I found myself at the Tick Tock in January of 2025, down from Vermont on business, nursing a cup of decaf and a bad case of insomnia and deciding I needed to make some changes in my life. The insomnia hit in middle age, eyes snapping awake at 1:00 a.m., maybe 2:00, with a grim clarity that there would be no sleep before 3:00, maybe 4:00. "The brother of death exacteth a third part of our lives," Sir Thomas Browne lamented in 1663. *If only.* Brother of death, come get it anytime.

Clocks confront you everywhere in the Tick Tock. Gleaming neon ones near the doors. An *O* turned into one on the cover of the menu. Stencils on the walls. But they are all sacrificial victims. The real promise of the Tick Tock is that the day's rhythms have been banished from its bright cocoon. You can get the waffles at 5:00 a.m. or 5:00 p.m. The lighting will be the same. When someone in your party wants the steak-house burger, someone wants the cheesecake pancakes, and someone just needs a martini, everyone's good.

Except we're not. My insomnia put me right in line with 50 million other American adults who suffer occasional or chronic sleep problems. There's no shortage of reasons. Stress, poor health, overstimulation, all the usual candidates. But for years, sleep experts have been shouting at the top of their lungs that part of the problem lies in our lightscape.

As I stirred my decaf and watched the lights of an ambulance go by, feeling the neon-lit loneliness of something that just crawled out of a Hopper painting, I was ready to listen. I'd spent a lot of time thinking about all the light we can't see—the infrareds and ultraviolets of the solar spectrum. But I'd spent almost no time considering the photons that have the most glaring impact on mind and body—the visible ones pouring into our eyes.

My server, smelling of onion rings, dropped off the check and told me to take my time. I didn't mention to her that shift workers suffer alarmingly high rates of diabetes, obesity, heart disease, digestive disorders, autoimmune disease, dementia, depression, and cancer. Or that night owls suffer more cognitive decline than early risers. Or that the WHO now classifies shift work as a probable carcinogen.

The biggest factor in this grim situation, say the experts, is artificial light at night, or ALAN as they call it. We did not evolve in Times Square. But you don't even need to be a night owl or to work the late shift to suffer the debilitating effects of ALAN. You practically can't avoid it.

For context, picture the world pre-Edison. The standard measurement for light is lux. A candle at arm's length produces 1 lux. Until the advent of artificial lighting, that was about as bright as night got for most people. The full moon produces a mere 0.1 lux. Even a raging campfire might throw only 30 or 40 lux.

By comparison, the light in the Tick Tock was around 200 lux. I knew because I had a light meter sitting on my table. It was shaped like a cheap plastic GPS device with a yellow body and a round light-sensing eye on top connected by a little neck. It felt intentionally anthropomorphic, C-3PO without the legs.

To be fair, 200 lux is not insanely bright. It's about twice that of a cozy living room and half that of a brightly lit office space. But it's about 199 lux more than the experts say you should have in your eyes while trying to sleep.

Rare is the bedroom that qualifies. More often, the technosphere seeps into your room, your eyes, your mind, with deadly results. In one recent study, volunteers who spent a night of sleep in 100 lux of ambient light felt that they had slept normally, and technically they had, but EEGs showed that they barely made it to the deep, restorative stages of sleep. Their heart rates stayed elevated through the night in a kind of superficial pseudo-sleep, and their insulin function was that of a prediabetic. Even scarier data was reported in 2025 by a team of Harvard cardiologists, who found a direct link between ALAN, brain stress, inflamed arteries, and heart disease. "When the brain perceives stress, it activates signals that can trigger an immune response and inflame the blood vessels," explained the lead scientist on the study at Massachusetts General Hospital. "Over time, this process can contribute to hardening of the arteries and increase the risk of heart attack and stroke." The effect was clear as day: "We found a nearly linear relationship between nighttime light and heart disease: the more night-light exposure, the higher the risk. Even modest increases in night-time light were linked with higher brain and artery stress."

The scariest data of all come from that big UK Biobank study that tracked the light exposure of eighty-nine thousand volunteers. Yes, lack of daylight was deadly, but ALAN was just as bad. It didn't matter if people were out and about in the evening or just trying to ignore the streetlight pouring into their bedroom, the ones burdened with the brightest nights were 21 percent more likely to die compared to those who experienced relatively dark nights.

The researchers estimated that those who got neither bright days nor dark nights were shaving five years off their lives. Again, cardiovascular disease was the top culprit, but ALAN also triggers inflammation and is associated with higher rates of cancer, Parkinson's, and cognitive impairment, including Alzheimer's. The bad news is that artificial light at night pervades the planet, polluting the dark hours of 80 percent of the world's population, and it grows worse every year.

As a card-carrying light evangelist, you'd think I of all people would have had my circadian act together by 2025. But it turns out to be surprisingly hard to do in the modern world. We are all Tick Tockers, tethered to laptops and light bulbs. Summer or winter, my evenings were spent beneath a cascade of artificial photons. And those extraordinarily efficient LEDs bathed my mammalian brain in an uncanny radiance unlike anything seen in the planet's previous 1.6 trillion nights. Times Square was everywhere.

And it had begun to catch up with me. Insomnia, a general malaise that arrived a little earlier each winter, blood pressure creeping upward, gluey cognition, chi clogged as ketchup.

In the modern world, we've come to think that this is just part and parcel of aging. Metabolism slows. Energy sags. Sleep sputters. But the burgeoning field of sleep studies suggests it doesn't have to be this way. If our circadian funk is a product of the flickering hellscape of the modern night, well, that's fixable, though you have to really want it.

Which I did. By the time I found myself tapping my mug for a refill in the neon glow of the Tick Tock, I was ready to admit that I was just another washed-up insomniac with a light problem. And it was time for an intervention.

CIRCADIAN, FROM THE LATIN *circa* (around) and *dies* (day), refers to the twenty-four-hour cycle of the spinning Earth and how we relate to it. There is a natural rise and fall to our hormone levels, alertness, and associated activity through the day, all governed by our built-in body clock, and we are best when it's in sync with the rising and falling of the light. Broadly, stuff gets done during the day, stuff gets fixed at night.

That goes down to the cellular level. Like tiny factories, cells spend their days burning energy and using it to do the thousands of different tasks they need to accomplish. All that work is physical, even if it's cognition. Molecules are made, broken, and remade. Electrons and protons are stripped apart to produce electrical current.

For their size, the voltages produced inside cells can rival that of a lightning bolt, and every day of activity results in a lot of molecular damage and waste products. Muscle proteins break. Neurons, firing hundreds of times per second, get clogged with their own gunk—and the accumulation of that gunk is what triggers the need to sleep.

Every evening, all this day work shuts down so the night repair crew can fix the damage, take out the trash, and top up the fuel supplies. As in a power plant, you can't really accomplish that kind of maintenance while the reactors are running. So it happens during sleep. Our cells go quiet. During a phase of sleep called slow-wave, the brain's vascular system pulses rhythmically, literally pumping cerebrospinal fluid through its innards and washing them clean. We can defer that maintenance for a while, but the more we do, the more junk and damage accumulates, and eventually things stop working so well.

Sleep is regulated by a hormone called melatonin, which is produced in the pineal gland and is triggered by special light receptors in our eyes. When light disappears in the evening, the pineal gland starts producing melatonin, the molecular messenger of darkness. Melatonin floods the brain and body, telling each cell that night has come. By daybreak melatonin levels are already declining in most of us, as the body clock anticipates the day to come, but the first splash of light in the eyes helps hasten the process and keep us in sync with the sun.

The receptors in the eye that regulate melatonin are particularly sensitive to light in the blue part of the spectrum. That's an ancient evolutionary adaptation to the color of the sky during the blue hour, that eggshell interval prized by artists. Blue is the first color we see at daybreak and the last to fade from the sky at night. In prehistoric times, that blue sensitivity would have meant that melatonin production fired up a couple of hours after sunset and shut down just before sunrise.

It still does in hunter-gatherer groups such as the Hadza of Africa's Rift Valley. By their own accounts, they sleep like a dream. These groups have no word for insomnia in their language, and when asked about it by anthropologists, they struggle to understand the concept. Anthropologists chalk this up to their "circadian hygiene." They know day from night.

We don't, as sleep researchers have been warning with increasing stridency, and we haven't since the arrival of electric lights in the late 1800s. Campfires and candles were warm and golden, with lots of red and orange and virtually no blue component, and even the gas lamps of the 1800s were essentially potted fire. But the "cold" light of electricity was something else and was not greeted with universal awe.

In his 1878 essay "A Plea for Gas Lamps," Robert Louis Stevenson makes it clear how weird the strange electrical illumination seemed. "A new sort of urban star now shines nightly," he lamented. "Horrible, unearthly, obnoxious to the human eye; a lamp for a nightmare! Such a light as this should shine only on murders and public crime, or along the corridors of lunatic asylums. To look at it only once is to fall in love with gas, which gives a warm domestic radiance fit to eat by."

Electric lighting might be fine for the harsher parts of town, Stevenson argued, but "where soft joys prevail, where people are convoked to pleasure and the philosopher looks on smiling and silent, where love and laughter and deifying wine abound, there, at least, let the old mild lustre shine upon the ways of man."

But it was not to be. The "old mild lustre" was soon relegated to séances and the occasional romantic dinner. Since then, our evenings have been brighter and our melatonin diminished. The problem was bad enough with the old incandescent bulbs, which produced only a little bit of blue light mixed with warmer tones, but it has become severe with the advent of modern LEDs, which are blue lightsabers, dicing melatonin to bits.

One classic study by Harvard's Charles Czeisler, a pioneer in the circadian field, compared an evening of reading a printed book versus reading on an iPad, which, like all modern devices, has a screen of tiny LEDs skewed heavily toward the blue spectrum. The subjects read from 6:00 to 10:00 p.m. in a dimly lit room of 3 lux, which is just enough light to read by. Light exposure at the participants' eyes measured 1 lux for the print group and 32 for the iPad. Over the course of the session, melatonin levels rose 19 percent in the print readers, as you'd expect as the evening wore on, but plunged 55 percent in the iPad group.

Media coverage of studies such as Czeisler's triggered a wave of panic about devices and their impact on sleep, and soon everyone was fitting their tablets with blue filters or donning blue-blocker glasses to prevent all those blue photons from reaching their eyeballs at night.

But it soon became apparent that screens were the tip of the blue iceberg. If you are reading a dimmed tablet in a dark room, few lux are reaching your eyes.* LED bulbs, on the other hand, are now ubiquitous in most homes, adding hundreds of lux of mostly blue light to our evenings. One recent study found that the average amount of ambient light it took to suppress melatonin levels by 50 percent was just 25 lux, the equivalent of a dim bedside lamp.

So it isn't just the night owls in the Tick Tock. We are all living an experiment to see what happens when you crush a society's melatonin.

I PAID MY TICK TOCK tab and ran the gauntlet of Times Square (78 lux in the dark recesses; 500 lux staring straight at a billboard) back toward my hotel. Had I come unstuck from the

* Some more recent studies indicate that the mentally arousing content of the reading may be a bigger concern than the light itself.

planet's rhythms? All winter, I'd been feeling it, kicking myself as I spent another dark day gazing into the abyss of the internet doing research, kicking myself some more as I lay awake at night, worrying that another evening's dose of photons was sabotaging sleep and feeding a festering mass of inflammaging.

There was one way to find out. I needed an LEDetox. A month of primeval darkness. Some place where the days were days and the nights fell with thudding certainty. Bonus points for weather nice enough to stay outside after dark.

The Rift Valley? Awesome, but impractical.

Rollier's Institute for Heliotherapy in Switzerland? Closed since the 1950s.

California, Sanatorium to the World? Times have changed.

Camping was also not an option as I had deadlines and needed enough infrastructure to keep my writing career on life support. Laptops and tents don't mix.

After much scouting, I found an off-the-grid 1962 Airstream for rent in the Sonoran Desert, plunked down on seventy-three hundred acres of nothingness an hour from the Mexican border. Remote, sunny, some of the darkest skies in the country. A solid stand-in for the Serengeti in the dry season. I booked it for all of March and vowed to never touch a light switch. The first step toward getting right with light, I decided, would be making friends with the dark.

CHAPTER 12

DARKNESS ON THE EDGE OF TOWN

Two Weeks of Night Without Light

Southwest of Tucson the land opened onto a wide plain too dry to support all but the most eccentric of life-forms. The light turned angular and the ranges ran like guardrails toward Mexico. The road cut straight toward the sky. In the middle distance, sandstorms churned across the plain.

I turned off the pavement and jounced along a washboard dirt road through miles of rangeland, my truck squirrelly in the sand. To the west, the setting sun backlit the strange monoliths of Kitt Peak National Observatory, set atop its seven-thousand-foot mountain like a modern-day Stonehenge. It was here for the same reason as me: because almost nobody else was.

There was a gate right where the directions said it would be, a cattle guard, a mile of ranch road through the thorn scrub, another gate, and then the Airstream, a chrome capsule crash-landed on a marginal planet. Behind it loomed Gunsight Mountain, a pink rock pile spiked with saguaros.

The Airstream showed its sixty-plus years, its shiny skin pockmarked with old injuries. Inside it was tricked out with Route 66 memorabilia and cactus Christmas lights. Outside were a few solar panels propped on cinder blocks, a firepit, and the nose cone of a practice bomb cadged from one of the many proving grounds that litter the Sonoran Desert.

A guy in a truck straight out of *The Road Warrior* pulled up to show me the ropes. He wore desert fatigues with a gun on his hip, a bowie knife, and a wraparound neckband Bluetooth phone that gave him a cyborg look. Deacon and his wife had bought the three-hundred-acre ranch during the pandemic and were trying to make a go of it, renting out the trailer and a couple of bare campsites. They had a well that produced a few gallons of water per hour, the difference around here between viability and not. He filled my water tank from a plastic bucket in the back of his truck and said he'd be back in a few days to check the level. "Normally people just stay here a night or two," he added. "But you're not normal."

Fair enough. I explained my project again so he wouldn't think I was a complete nutter. Light, insomnia, LEDs, yada yada yada. No light after dark. All month. I picked his place because of the location and the rave reviews of the stargazing.

Deacon nodded. Didn't they have light switches where I live?

Well, yes, but . . . better here. Stargazing in March snowstorms has limited appeal. Also, my partner had no interest in a month of darkness.

Deacon shrugged. Behind him, Venus, the crescent moon, and the mountains were conspiring to create something that should have hung in the Louvre, but he'd seen a thousand before. He was more intrigued by the notion that his land might have a particular appeal to light refugees fleeing urban hellholes. I was welcome to roam all three hundred acres, plus the adjacent seven thousand acres of public land. There were potsherds and petroglyphs all over, plus an old adobe cabin up in the arroyo. Just watch out for the meth heads. More important, watch out for the rattlesnakes. They are going to wake up any day and you are pretty much standing in Rattlesnake Central and they are hard to see. Also, don't make a fire in this wind. It hasn't rained in a year. The whole place is a tinderbox. Other than that, have fun.

Deacon clattered off in his truck, leaving me to contemplate my new home. Venus was already gone, the moon, too. The stars were as advertised. I whipped out Threepio. The evening was already down to 0.1 lux, just one-tenth of a candle flame. Sensational darkness.

But I was already quivering as the wind sucked the heat out of me. Doesn't it get to be 109 degrees here? How could it also be 38? There would be no leisurely decompression beneath the vault of heaven tonight. I retreated into the trailer, pulled on my warmest clothes, and huddled under my one blanket. Sleep was not in the cards.

But that was fine; I'd given myself one night to get accommodated. Tomorrow I had a date in Tucson at the University of Arizona's Center for Sleep and Circadian Sciences, where a postdoc named Kat Kennedy had promised to guide me through the land of Nod and help me with my protocol for the month. After that, no more excuses.

Through the black night, the wind raged, hissing through the cracks in the old Airstream, rattling the windows, rocking it on its tires. The whole thing felt as if it might launch like Dorothy's house and go spinning through the skies, crushing someone in Sedona. Circadian alignment felt far away. But I had found the dark.

THE CENTER FOR SLEEP and Circadian Sciences is a $5 million warren of corridors, guest suites, and control rooms built into the basement beneath the U of A hospital. Light, sound, temperature, and even oxygen concentrations can be precisely controlled. Each suite has a cushy bed, ports in the wall for various diagnostic devices, cameras and intercoms, and spacey light panels. It felt like a cross between a Comfort Inn and a panic room.

Kat Kennedy gave me a quick tour so I'd know what I was in for. The high-tech LED panels in the ceiling, made by a Bay Area start-up called Telelumen, were capable of creating virtually any dynamic lighting scheme imaginable. "It's very niche," she said. "The new software allows us to get really granular, millisecond to millisecond."

The sleep center uses the lights to study people's response to different wavelengths, intensities, and patterns. At my request, Kat had agreed to put me through twenty-four hours of different lighting schemes in one of the sleep suites so I could get a sense of how particular lightscapes can alter consciousness and alertness. The plan was for me to spend two weeks in the desert, wearing an Actiwatch and keeping a sleep diary, then come to town, turn in my data, check into my room, and trip the light fantastic. If that magical mystery tour actually resulted in great sleep, so much the better.

Kat had long, wavy hair and limb-length tattoos. She spoke in soothing British tones that must be a balm for anyone suffering from serious sleep disorders. Like most sleep researchers, she thought daylight saving time should be abolished.* She was born in Egypt but raised in the UK, which never suited her. "Too crowded, too dark, too rainy," she said. "I really struggled with the lack of sunshine. I definitely have a high risk of seasonal affective disorder." In her twenties, she was beset with vivid dreams of Southwestern

* Perhaps not coincidentally, she lives in the right state for it. Arizona ditched daylight saving time in 1968.

landscapes, and she followed them here. "I can't explain it," she said. "It just felt like it was calling. I always say I was born in the desert and I will die in the desert."

Her sensitivities led her into the field of circadian rhythms. Her specialty is "mind after midnight," the fragile state of the late-night brain, when, as she put it, "our decisions are not guided by our best self." With the rational mind down for maintenance, anxiety spirals and rates of violence and self-harm rise sharply. For people forced by physiology or duty to be awake, she researches better ways to cope. "I always think of sleep as being a good gauge of how society is doing," she said.

How are we doing? Not so good. One-third of us don't get the seven hours of sleep we're supposed to. The cost to society is a staggering $400–500 billion due to disease, accidents, and lower productivity. And while the problem plagues all demographics, it's particularly worrying in a group that used to be somewhat protected: kids.

We popped into the office of Kat's colleague Lauren Hartstein, who had an Actiwatch left over from a previous experiment that I could borrow to track my sleep and light exposure for the next two weeks. Hartstein is one of the few researchers in the world to specialize in light's effect on preschoolers. "There's a lot of focus on teens and adults, but not as much on little kids," she told me. "But preschoolers are going through a huge change. They're setting up the habits they're going to carry through childhood." About a quarter of kids develop sleep issues, and many of them are now dependent on melatonin "sleep candy" in a variety of fruity flavors.

Hartstein got interested in light's impact on kids a decade ago after multiple studies showed that learning was greatly improved by sunlight. Students scored up to 25 percent higher on standardized tests administered in daylit rooms compared to rooms using artificial lighting.

For her dissertation at the University of Massachusetts, Hartstein found that preschool kids performed cognitive tasks significantly better under daylight-mimicking "bright" LED bulbs, which produce a lot of blue light, than under "warm" bulbs of the same intensity that produce more red.

That was in 2017, and Hartstein was ahead of the curve. "In fact," she said, "when I brought it up to my PhD adviser, who was a classical developmentalist who was getting ready to retire, he was like, 'Okay, I guess we can do this project, but half an hour of light is not going to do anything to these kids.' And then when we found all these effects, he was shocked, and he eventually came around to the whole thing."

Light's impact on preschoolers' sleep turned out to be such an important issue that it quickly became Hartstein's focus. "We found that kids are just absurdly sensitive to light exposure before bed. Even really low intensities suppressed their melatonin and shifted the timing of their clock almost an hour."

The results were the kind of thing to, well, keep a parent up at night. An hour of dim light of just 5–40 lux was enough to suppress kids' melatonin by 78 percent. More light suppressed it even more. "These are quantities of light that wouldn't be setting off adults," Hartstein said. For most kids, the impact was still being felt an hour after the light was switched off.

It's all enough to make an insomniac twitch, but Kat pointed out that the solutions are straightforward. "The body knows how to sleep if you create the conditions for it to do so." In her own life, she actively cultivates those conditions. She rises at sunrise, forgoes breakfast for at least an hour to let the melatonin fade from her bloodstream, Pelotons every morning, eats her meals at the same time every day. At night, she employs a suite of tools to shield herself from light: an eye mask for sleeping, blue-blocker glasses for watching TV, adjustable smart bulbs to shift toward warmer, more soothing colors in the evening.

"I live at a very busy intersection right next to the university, so I do what I can," Kat said. "Ideally I would live somewhere quieter with less light pollution and fewer ambulances blaring past, but it's what I can afford and it's convenient for work, especially when I have to be in the lab at weird hours." One of the ironies of being a sleep researcher is that you sometimes have to pull all-nighters with your subjects.

But across society, light at night has become so common that few of us have experienced the full stop of real darkness or even know what we're missing.

THE SONORAN DESERT TURNED out to be a pretty good stand-in for the Rift Valley. Mesquite trees for acacias, pronghorns for gazelles, meth heads for leopards. It was just a few days from the spring equinox, so the day length was the same planetwide, twelve and twelve, cosmic balance.

Not that I'd be making full use of it. As I sipped coffee on a rock and watched a family of Montezuma quail scuttle around Deacon's slapdash solar array, picking at dried seeds, I tried to make my peace with the knowledge that I was going to be letting some spectacularly sunny days pass me by. If I wanted to isolate the effect of simply eliminating light at night, that had to be the only variable I changed. And that meant spending my days as I usually do: in front of a laptop.

Realistically, I needed to do that anyway. Life being what it is, I hadn't quite managed the clean break from work I'd hoped for. So the plan was to spend the next two weeks working as usual in the trailer, with evenings outside, but not a drop of light after dark. Then I'd report to the Sleep Lab for twenty-four hours of observation and experiments, and depending on how things were going, Kat and I could adjust the plan for the final two weeks. Deacon was pleased to know that I was getting my money's worth out of the solar power and satellite internet he'd gone to such trouble to install.

What about the blue laptop screen? Not a problem during the day. Here on the rock, the vault of heaven was pouring a million times more blue photons into my eyes than my measly MacBook could ever muster. Besides, blue's not evil. It makes sense to avoid it in the evening, but during the day, it helps suppress melatonin and boost cortisol, sharpening awareness and mobilizing energy reserves, which is exactly what you want. If anything, I needed more blue than I got during my day in the crepuscular light of the trailer.

In early evening, however, I climbed out of my capsule and began to explore the shimmering, parched planet. The wind had lain down at last. Sandy washes wended around rocky islets studded with purple prickly pears. It felt like a coral reef that had lost its ocean. Even the saguaros had the stupor of passengers on a long-haul bus ride.

I found the adobe cabin up in the arroyo next to a broken cistern. The mattresses and tin cups spoke to an unplanned retreat. Ditto the bleached cow bones in the corral. Vultures pirouetted on the thermals.

As the sun crept toward Kitt Peak, I gathered wood for a fire, impaling myself on sneaky cholla buds. Over the next two weeks I would turn into a pincushion, but the desert is God's gift to the half-assed fire maker. I chose from a lifetime supply of mesquite so light and crackly it felt kiln dried. I didn't even need paper. The dried grasses went up like Roman candles. I cracked a can of beer, got the grill good and hot, blackened a bunch of vegetables, rolled them into a tortilla with a little hot sauce, and ate standing up, bits of filling slipping out the back of the tortilla and falling on the coals with a hiss.

It felt strangely radical to have the sunset be the entire agenda. It passed through a dozen stages, from fiery orange to astral lavender.

All the time Threepio slowly clocked down, hundreds of lux, then dozens, then just a few. A coyote sent the sun on its way with a long salute.

Finally the color was gone except for some peaceful blue bands along the skyline. It wasn't even dark, and I already felt different. Instead of compensating for the dying light by flicking on the house lights, then keeping them steady until lights-out at eleven, I liked the slow fade, my mind decompressing like a leaky air mattress.

As the fire faded to coals, a thousand eyes opened above me. The temperature was dropping fast, so I hauled my sleeping bag out on the sand and squirmed inside. Venus set, chasing the sun, the crescent moon right behind. Jupiter hung huge between Orion and the Pleiades, Mars an unmistakable red dot in Gemini.

My eyes kept adjusting to the deepening darkness. We are almost never in the dark long enough to discover how much better our night vision can get. I picked out the faintest star I could find in the darkest quadrant. Unlike the billions of photons per second from the sun, perhaps a dozen per second were trickling into my retina from this distant messenger. When they departed their star years ago, the odds that they would end their journey inside a human eyeball were remote indeed. I scanned the sky, enjoying the knowledge that I was acquiring iotas of energy from across the galaxy.

It was all appropriately cosmic, and yet the arc of stargazing bends toward boredom. Soon my eyelids were slamming down. I congratulated myself on such a spectacular plan and headed for the Airstream, gleaming coldly in the starlight.

But when I opened the door, my rookie mistakes came clear. I couldn't see a thing. I felt my way blindly through the unfamiliar space, from table to sink to bathroom door. Where was my sweatshirt? No idea. How about my toothbrush? I felt my way around the bathroom and knocked over the plastic cup I'd put it in. Oh well.

I made some mental notes to put everything in its proper place tomorrow and to be sure to complete any essential tasks before the shroud of darkness fell. Then I pulled my way back to the bed on the other end of the trailer, pushed aside all the clothes that I hadn't yet stored, lay down, and conked out.

But when my eyes snapped open in the dark, I could tell it was still deepest night. Only a few hours had passed. I could also tell that I was coming fully out of sleep, a feeling I knew all too well. My hopes sagged. Sure enough, I was still awake hours later when the sky began to blush. I decided to just get up and get on with it.

I hoped for better the next night, and the next, but I didn't get it. I stuck with my routine. Sunset, campfire, dinner, stargazing, sleep. That part went brilliantly. My body couldn't wait to shut down for the day, and the initial sleep was sumptuous. But it never lasted past the wee hours. I tossed and stewed in the useless dark, the Actiwatch on my wrist silently judging me.

As one week stretched into two, there was no shortage of time to mull the details. Clearly artificial light at night was not the only root of my insomnia or these ridiculously dark nights would be curing it.

But the experiment wasn't a total loss. I'd come to love the slow-drip evenings under the stars, the dreamy drop into a sea of melatonin. After a lifetime spent using evenings to go out or to catch up on reading, it felt like an incredible luxury to have time to put the whole cognitive house in order before closing it up for the season. I felt less frazzled.

Still, no matter how good your sleep is, if it lasts only a few hours, you are definitely not waking up with the house in order. It was time to consult the experts.

CHAPTER 13

THE LIGHT FANTASTIC

What I Learned in Sleep Lab

On a bright March morning a quarter year from my lonely night at the Tick Tock Diner, I hopped in my truck and drove east on the Ajo Highway toward Tucson and the Sleep Lab, the early light splashing across the truck hood into my tired eyes. For two weeks, I'd been ALAN-free. That should have been more than enough time to reset my circadian clock, which should have responded to a new light routine within a few days. Clearly it wasn't that simple. I looked forward to a change of pace, however weird. Anything but the Airstream.

Kat Kennedy met me at the hospital entrance and whisked me down to the basement. Some of the rooms were filled with volunteers in a study to determine how sleep deprivation affected appetite, but the door to one of the suites had a sticky note on it that read KAT'S PARTICIPANT.

After the brilliant light of the desert, the soft-lit room felt eerie as a partial eclipse. The walls and bedspread were a calming Weimaraner blue. The desk and cabinet IKEA plain. The place was Spartan, with just a granola bar and a couple of bottles of water for sustenance. I'd be stuck in this room for the next twenty-four hours, and I'd been told to bring any food I'd need, so I'd picked up some tamales on my way in. That would have to get me through.

I had no idea how long this particular lighting scheme would last, or what would follow. Ostensibly the goal was for Kat to use the fancy lights to manipulate my level of alertness through the day in a way that might lead to a good night's sleep, but I figured this was likely to be my one and only experience of a state-of-the-art sleep lab, so I hoped she'd make it interesting. For her, it was a good opportunity to test out the new lights on a pliant volunteer.

We both thought my responses would be more telling if I didn't know what was coming, and Kat wanted to get some data by testing me throughout the day, starting immediately. She stuck a smartphone in my hand and fired up a psychomotor vigilance app. Every time a red stopwatch popped up on the screen and started counting milliseconds, I had to tap it as quickly as possible. The tests are widely used to measure alertness, and she was curious how mine would change over the day. I pressed start and the stopwatch started bouncing around the screen and counting. I chased it, tapping wildly. My first efforts were pathetic—419 milliseconds, then 500—but soon I started to get a rhythm and whittled it down to the 300s.

The test ended and Kat nestled a black plastic bar shaped like a bighorn sheep rack over my head to get an EEG brain scan. I felt the prickles as its electrodes locked onto my scalp. Kat told me to sit staring at a spot on the wall for three minutes, then to close my eyes and relax for two more minutes. She removed the rack and handed me pen and paper for a word-fluency test. "Write all the words you can think of that begin with *F*. You have one minute. Go!"

I dredged up my old Boggle tricks. *Fun, funny, funnily. Fan, fin, fen.*

My testing done, Kat collected my Actiwatch for data retrieval. She'd be back in an hour for another round of tests. In the meantime, I could settle in. The lighting would take care of itself, don't touch anything, and also there's an intercom on the wall and a camera in the ceiling, so you know, best behavior.

She slipped out and I flopped down on the bed, jarred by the unexpected testing. There wasn't much settling to do. I stuck my tamales in the minifridge and acquired the WiFi password for my laptop. The whole thing felt strangely luxurious after the Airstream, the shower looming large. But in the soft light I felt hazy. I consulted Threepio: 33 sad little lux. Did they want me sleeping all day? I tried to work but it was hard to focus. My next tests showed that there wasn't a whole lot of vigilance in my psychomotors.

At 11:57 a.m. the lights suddenly surged from Partial Eclipse to Hospital Harsh. The charm was routed from the chamber, the room saturated in 555 lux of icy whiteness. The walls weren't Weimaraner blue after all; more cactus gray. Fully awake at last, I ripped through a few hours of long-neglected emails and crushed my next vigilance test, whittling my times closer to 200 milliseconds. Kat ran me through another EEG session and Boggle word riff.

Through the afternoon, I hit a groove of productivity. It all ended at 3:58 p.m., when the lights turned Shrek green. 62 lux. It was so unexpected that at first it felt as if the day had banked into a woozy barrel roll, but I soon adjusted and began to feel almost giddy with the absurdity of the situation. I might as well have been inside a margarita. I got little accomplished and didn't much care.

At 5:00 p.m. Kat popped back in for a final EEG reading while I stared at the now-green wall. And that was it. She was going home in an hour. There were sleep technicians on call all night, so if I needed anything, I should just shout through the intercom or jump around in front of the camera. No eye masks! Good luck, pleasant dreams, see you at 8:00 a.m. tomorrow.

And then I was all alone in Sleep Lab, the long night stretching uncertainly before me. The extragreen tamales went down with effort at 7:00 p.m. Around 8:00, the lights turned Amsterdam red. The veins in my arms showed black and prominent. It was intriguing, but soon my optical system was glitching badly. When I put my hands over my eyes, I saw flickering blue webs of chaos. The red glow was just 66 lux, but it was a weird sort of smoldering hell. I popped in my headphones and tried to lose myself in a Cormac McCarthy audiobook.

At 9:00 I was still awake. Somewhere in the desert, nighthawks were rising from reefs of bloodred clouds. At 10:00 I'd completely lost the thread of the story. By 11:00 scalps were being taken and the unsettling thought arose that Kat had decided to sear the insomnia out of me with a red-light all-nighter. That dread had blossomed by midnight, when the room suddenly went pitch-black and I developed a minor case of Stockholm syndrome.

And that's the last thing I remember.

AT 4:48 A.M., a soft 1 lux glimmer suffused the chamber. It took a few minutes for me to confirm it wasn't my imagination. Over the next hour the light slowly rose to the same soothing 33 lux blue-gray it had been when I arrived. I tipped an imaginary hat in Kat's direction. I couldn't say I was well rested, but at least the dreadful pattern had been broken. Nearly five solid hours.

But now, novelty gone, it started to feel as if I were trapped in a flat afterlife. Kat's 8:00 a.m. arrival felt eons away. I killed some time reading studies on daylighting and hospital stays. A Korean study in a large general hospital compared average length of stay for patients in rooms that faced either southeast or northwest. The southeast-facing rooms received twice as much sunlight as the northwest-facing rooms, and patients in those rooms recovered 16–31 percent faster. Another Korean study of eighty thousand hospitalized patients found that those with a bed near the window had significantly shorter lengths of stay than those with a bed near the door, where they received almost no natural light.

I myself would have liked to be released 31 percent faster. While I waited, I kept reading. Researchers in Copenhagen measured the light coming through the windows of a psychiatric ward in each season. At noon on the winter solstice, the northwest windows were getting a meager 1,200 lux, versus 20,000 through the southeast windows. At the equinox, it was 2,000 versus 40,000. Even on the summer solstice, a feeble 3,000 lux was passing through the northwest windows, while a near-outdoors-worthy 60,000 lux was streaming in the southeast glass. Among patients hospitalized with severe depression, those in southeast-facing rooms recovered in just twenty-nine days, versus fifty-nine days for those in the northwest-facing rooms.

The effect isn't limited to people with severe depression, either. Sunlight is often found to be just as effective as antidepressants in relieving depression and seasonal affective disorder. It boggles the mind that we aren't incorporating it into hospitals, nursing homes, prisons, and other artificially lit facilities.

I was clawing at the walls by the time Kat rapped on my door at 8:00 a.m. sharp. "Would you like to debrief here or somewhere outside?"

It was not a hard choice. The sunlight felt extrasweet, the shadows still long across the campus. We found a table and Kat showed her cards.

The dim light so early this morning was to mimic the desert dawn. "One of the things we preach with sleep hygiene is to keep your morning consistent. I knew you'd be going straight back to the Airstream, so I wanted to keep you on that same type of morning schedule." But she'd actually one-upped the dawn by starting things an hour early. "I thought that the best chance of you having a consolidated bout of sleep was to push the bedtime later and the morning slightly earlier, so the sunrise simulation started at five a.m."

The strong full-spectrum light midday was intended to enhance my cognitive performance. It takes strong light—stronger than most indoor settings provide—to fully boost alertness, which is why you can spend hours in the morning stumbling about the house in a fog if you don't step outside. For me, the bright white light worked: My scores on both the vigilance and word tests were significantly better than in the soft morning light.

The green light was a curveball. A decade ago, the Harvard neuroscientist Rami Burstein discovered that most wavelengths of light made migraines worse, but green light unexpectedly relieved them. We have three different types of color receptors in our eyes, tuned to red, green, and blue photons. For whatever reason, the green receptors send smaller, less stimulating electrical signals to the brain, so green light feels less oppressive.

Green light also reduces neuroinflammation and triggers a release of natural opioids, promoting a serene state that was obvious in Burstein's patients. "Almost unanimously, they describe green light as calming, soothing, relaxing," he has said. "And after enough time in green light, they talk about the fact that their anxiety gave way to feeling calm and their irritability gave way to feeling relaxed." My green-phase EEG showed a big spike in the low-frequency brain waves associated with meditation and creativity. Kat's colleague Mohab Ibrahim now uses green light to treat migraines, fibromyalgia, neuropathy, and anxiety.

Because our melatonin response is tuned to blue light in particular, the farther a wavelength is from blue, the less impact it has on melatonin. Green has only a small effect on melatonin, and red even less. The goal had been to keep me awake but unstimulated so that when lights-out finally came, I was primed for deep sleep.

For someone with sleep-onset insomnia, Kat said, lights-out would have come earlier, and the scheme would have been different. "I wouldn't have used the crazy red light. Just a mixed spectrum, but low lux. And I wouldn't have had the dim light in the morning and the bright light in the afternoon. I would have flipped that, progressively going from bright light to dimmer."

Kat hoped that the introduction of more dynamic lighting systems such as the Telelumen could lead to healthier indoor environments. "There are a lot of possibilities," she said. "One of the things I'd like to play around with is trying to mimic the sun passing overhead, moonrise, things like that."

It was an intriguing idea. I pictured homes, offices, hotels, where the light plays all day, keeping people on the circadian path. A few years ago, the light pioneer George Brainard led a project to retrofit the lighting on the International Space Station to simulate sunrise, midday, and sunset to improve astronauts' sleep.

But the reality is that no artificial lighting system, no matter how sophisticated, is ever going to be able to hold a candle to natural daylight. Although Kat thinks bespoke LEDs can be a valuable circadian tool, she said the best sleep aid is bright daylight, because of the contrast it makes with nights, even if they aren't superdark. But these highs and lows are often missing in the modern cityscape. "That dampening of amplitude means we have weaker signals going to the brain," she said. "I think that's where circadian alignment can really help."

It's not just the overall light level that tells your body to rest, it's also the magnitude of the drop. "Try and maximize that daytime signal so that when it's nighttime, you have a better chance."

She pointed to some recent studies by David Samson, a University of Toronto anthropologist, who tracked the sleep patterns of hunter-gatherers using Actiwatches. Sure, they don't have a word for insomnia, and they say they sleep great, but check the data and it turns out they sleep just over six hours a night. That's forty-five minutes *less* than that of members of modern industrial societies, who average about seven.

But apparently that's plenty. They rise before dawn and spend all day outside. That, Samson and others believe, is the real secret to their good health, and our bad. "The key issue in economically developed, industrial economies is not sleep duration but primarily circadian disruption," Samson wrote in the *Proceedings of the Royal Society B*. "The very factor that may be improving sleep (e.g. secure, regulated sleep sites) may also be masking access to circadian entrainment—such as sunlight during the day."

In other words, our cozy, cloistered homes and bedrooms are partly responsible for our sluggish ways. That was also the gist of a recent study at the University of Colorado. Researchers fitted eight healthy young adults with Actiwatches and tracked their sleep and light exposure for a week during their regular lives, then made them go camping without artificial lights for a week. Their average light exposure during their waking hours while camping was more than four times higher than at home, and the amount of time they were exposed to at least 1,000 lux, which is a good amount to guarantee full alertness, was eleven hours per day while camping versus just four at home. (But even that was a vast improvement on Chicago office workers, who managed just two hours above 1,000 lux in summer and a demoralizing one hour in winter.) Overall, the dark nights and bright days shifted the campers' sleep patterns one to two hours earlier, closer to the solar schedule.

That shift pays off in better health and productivity. We can even put numbers to it by looking at time zones. Time zones are an attempt to keep everybody in the world aligned with the sun, but they are clunky tools. Each of the twenty-four zones is about one hour wide, which means there's an hour of difference between when the rising sun hits the eastern edge of the zone and the western.

For instance, on the spring equinox, the sun rises in Boston at 6:45 a.m., but it doesn't rise in Indianapolis, all the way on the other end of the same time zone, until 7:46. If people in both cities have to wake up at 7:00 a.m. to start their day, the Bostonian is going to be awoken by sunshine in the eyes while the Indianapolitan is struggling to rise forty-five minutes before dawn.

At the end of the day, the problem is reversed, but still in favor of the easterners. Darkness arrives an hour earlier for them, giving their systems extra time to relax, make melatonin, and improve the odds of an earlier bedtime and a full night's sleep.

So people on the eastern end of a time zone tend to have circadian rhythms that are more aligned with the day/night cycle. This shows up in the cancer rates, according to the National Cancer Institute. For every twenty minutes of later sunrise farther west in a time zone, total cancer rates rise by 3 to 4 percent. Another study found that people on the wrong end of the time zone averaged nineteen minutes less sleep and earned 3 percent less than their eastern-end compadres, presumably due to worse health and performance.

My marching orders seemed clear. I'd been doing great work on the dark nights, but that was only half the cosmic battle. I hightailed it out of Tucson with a new rallying cry: amplitude. Dawn in the thorn scrub. Noon on a mountaintop. Night in the void. Brain wrung like a sponge every rinse cycle.

And light. Shifting, dynamic sunlight, the full multivitamin. Paying close attention to light's role in my consciousness had left me thinking about photons as building blocks of mind. And it had taken just one day of deprivation to leave me pining for the desert.

CHAPTER 14

TAKING THE CURE

Bright Days, Dark Nights, and Sleep at Last

The blush over the eastern ridge of Gunsight Mountain started early. There was a glimmer at 5:15 a.m., a confirmed glow by 5:30. It got bigger and brighter and then sparked, orange plasma bubbling over the rim like lava, resolving itself into an apse and then a disk and then you couldn't look anymore. Ninety-six thousand lux! Goodbye, melatonin.

Goodbye, literary life, too. I had seen the enemy, and it was MacBook Air. The laptop got banished to the drawer for the rest of my stay. Clearly that was no long-term life solution, but I'd figure that out later. I had to take this circadian rehab one step at a time, and right now those steps led away from the trailer.

The Sonoran Desert was doing its part. Not even a wisp of cloud to dampen the extreme circadian amplitude. I already had my sleeves rolled up, sitting on my favorite rock, watching the last shadows retreat into the mountain. After two weeks of astronaut life in the tin can, the morning light on my bare arms felt as luxurious as a spa.

By then, I'd boned up enough to know that morning light has some exceptional qualities. Most obviously, the early light in my eyes was going to wake me up and sync my master clock with the solar cycle. And it was going to do so without causing damage. When it is low on the horizon, sunlight travels through five times as much atmosphere as it does when directly overhead, and all that atmosphere filters out the shorter wavelengths. So the light that reaches us has lots of infrared, red, and orange (which is why the rising sun looks red or orange) but little ultraviolet.

You can tell this by looking at the UV index, a tool that was introduced in the 1990s to help people gauge how much protection they need. For instance, on that bluebird March day in southwest Arizona, the UV index was supposed to peak at 8 at midday. That's a lot of UV. A UV index below 3 is considered negligible, while 3–5 is moderate and people with fair skin would require protection for extended times spent outside. Anything higher than 6 is pretty intense for all but the darkest skin tones.

So I'd definitely want protection by late morning. But at that moment, the UV index was only 0.1, and it wouldn't reach 3 until after 10:00 a.m. In the afternoon, it would drop to 4 after 4:00 p.m., and by 6:00 p.m. it would be less than 1. So even in a high-sun environment, morning and late afternoon can be golden.

Watching the orange light fall across the knobby mesquite trees, I thought again of prehistory, of how our days once began with this wash of low-energy photons over naked skin. Most of what we know about the damaging effects of UV comes from experiments in a lab, where skin cells are exposed to sudden blasts of pure UV. Over the years, various scientists have noted that this is not an accurate reflection of how things happen in nature, when UV arrives as a package, bundled with all the other wavelengths and preceded by an early-morning dose of red and infrared. Might that make a difference?

To find out, a Parisian research team first exposed human skin cells to a jolt of UV big enough to kill half of them. Then they followed that experiment with a second in which they prepped cells ahead of time with thirty minutes of infrared light before administering the same jolt of UV. Only a small percentage died. When they doubled the infrared pretreatment, none of the cells died.

The pretreatment caused the cells to mobilize their natural repair mechanisms. When the UV hit, they were ready. All damage was quickly and expertly fixed, and the cells were back to normal the next day. "The protection induced by infrared radiation described here seems to be a very effective mechanism of cellular defense against the cytotoxicity of solar UV, selected by and conserved through evolution, and thus of major importance for life protection," the researchers concluded.

A related experiment by the McGill University dermatologist Daniel Barolet found that pretreating fair-skinned volunteers with deep-red LED light was as effective at preventing sunburn as SPF 15 sunscreen. "This could suggest that mammalians already possess a natural mechanism which, in reaction to morning infrared radiation, prepares the skin for upcoming noon potentially damaging UV radiation," he wrote.

This implies that part of our modern problem with skin cancer is lifestyle, and it may help to explain why office workers are more susceptible to melanoma than outdoor workers. When we step straight from a dim building into blazing sun or, worse, straight from the depths of winter onto Miami Beach, we catch our skin cells off guard.

Out of curiosity, I decided to test the pretreatment theory, making sure my skin was alerted to each coming day by a dawn session of gentle light. (And since infrared light passes easily through clothing, I could do this even when the dawn was too chilly for bare skin.) Barolet has also found that blue, red, and near-infrared light are able to mobilize nitric oxide in the skin and reduce blood pressure, though not as effectively as ultraviolet light. Still, I'd take it as a bonus.

Pretreatment complete, I slipped on some hiking shoes, grabbed my pack and water, and headed for the hills. It was time to bag this mountain and wake *Homo perspirus* from his winter slumber.

The route was no route at all. Just rock over rock, a phantom rattlesnake curled in every crevice. The morning light cast jagged shadows across the pink stone. It caught the saguaro spines and illuminated them like desert saints. I was the first thing weighing more than a mule deer to come this way in God knows how long, maybe ever, and the cobble crumbled beneath me as I climbed.

Near the ridgeline the route got dicier. The folds of rock turned vertical, knifelike, a stegosaurus spine of giant sandstone plates. The sunbaked air smelled like creosote. I shimmied through chutes, semi-lost, scrabbling upward until my concentration and my fingernails were worn to nubs, and then suddenly, unexpectedly, I was on top.

But not first after all. Somebody had carved a perfect petroglyph of a sun into the sandstone. Not ancient, I assumed; just another enthusiast. Even so, it felt like a message.

Space fell away on all sides. Kitt Peak to the west. A quarter moon in the east. Far below, the Airstream gleamed absurdly against the ocher land. The only thing moving was a dune buggy cutting through the desert twenty miles away, kicking up a column of dust that hung in the air. I stood beside the petroglyph, an antenna atop the world. For the first time in ages, I felt in sync with the interlocking circadian spheres around me, a living Antikythera.

From the top it was easy for me to follow the ridgeline a few miles south, picking my way through gaunt ocotillo thickets. *Ocotillo* means "little torch"; hummingbirds, newly arrived from Mexico, buzzed the jets of red flowers at the plants' tips.

The day quickly got hot and I rolled down my sleeves and put on my hat. Hunter-gatherers tend to escape the heat with a midday siesta in the shade, and so did I, leaning against a cool rock wall on the northern side of the ridge. The moment had an addictive quality. It was the emptiness and the heat and the slow trickle of endorphins from the skin to the mind, the sense of being nothing but a nexus of energies caught between the morning desert and the night desert. Midafternoon, I started up again, following a circuitous route through the valley, which was a sheet of golden grasses beneath the black mesquite trunks. By evening I was back at the trailer.

Dinner tasted better than ever that night, my one can of beer close to ecstasy. The decompression around the fire felt like a vertiginous slide into oblivion. For the first time in months, I slept straight through. When I woke up, my neurons felt squeaky-clean.

That marked the beginning of a glorious run of days. I caught every sunrise. I bagged every peak within reach. And I slept. The days blurred together. I'd see the sun down, and the next thing I knew it was sneaking up behind me. It was like playing tetherball with fire.

I came to adore the arc of the day. The way the desert seemed to be holding its breath before sunrise. The charge of morning photons showering my skin. The sundial shadows of the mountains. Part of the pleasure was in matching my own metabolism to the light's, days of movement followed by nights of nothingness. I never did burn. I was careful to cover up midday, and the rest took care of itself.

By the end of the month, five pounds of *something* had melted off my frame, maybe water. Blood pressure reverted to normal. Chi flowed. My skin felt tighter. My mood turned gossamer light. My phone stopped recognizing my fingerprint. In the mirror I caught sight of a *Dune* Fremen, pinpoint eyes full of spice.

BY THE TIME I headed back to New England, I'd stopped thinking of the skin as a mere barrier between us and the world. It's more of an observatory, taking stock of the light and atoms that smack into us and making reports. The cross talk with the brain never stops. Neither does the flood of signaling molecules alerting the body to the latest developments. If we keep our skin in the dark, well, we keep it in the dark.

That's become even clearer thanks to several new studies, including one very large one using 339,000 UK Biobank participants, that found strong associations between vitamin D and the expression of many genes involved with the circadian clock and seasonal variations in metabolism. This was, the paper pointed out, "the first evidence to support a mechanistic role for vitamin D in human seasonality."

If that pans out, then it brings the vitamin D story full circle. Just as melatonin is the "messenger of darkness," vitamin D may be the "messenger of daylight." Produced by sunlight in the skin, it then visits individual cells throughout the body, keeping them in tune with the day and the season by helping to regulate everything from the body clock to energy levels, fat storage, immunity, and cognition. Blocking its production might be a very bad idea—and so might delivering a big slug of oral vitamin D first thing in the morning.

I tried to picture how it all worked back in our hunter-gatherer heyday. Morning sunlight fell on skin that was coated in microbes that had been flourishing on a diet of sweat, oil, and sunshine. The microbes started exchanging molecules and messages with the skin cells. Metabolism rose, inflammation receded, and the skin stocked itself with antioxidants and melanin in preparation for another action-packed day on the savanna.

By late morning, UV light was cascading out of the tropical skies. A small amount bypassed the microbes and melanin and caused some minor damage that quickly got neutralized by antioxidants and cellular repair mechanisms, which were at the top of their game because they'd been doing this every day of their lives. Meanwhile, the immune system was listening in on the cross talk between microbes, skin, and the rest of the body and concluding that there were no emergencies to get worked up about, instead quietly going about its business of mopping up stray inflammation.

It was, in other words, a normal day. Nitric oxide flowed. Blood vessels relaxed. Endorphins percolated. Energy and alertness surged, then dissipated with the coming of night. Life got lived.

Then I tried to picture the confusion of these same systems experiencing a typical modern day. The skin microbiome is different, composed primarily of species that can survive the chemical environment of modern skin-care products. The skin is pale, there being no call for melanin. Lack of sunlight and exercise mean damage-repair systems haven't been tested in ages. Vitamin D is deficient. Nitric oxide, too. Inflammation rises. Mood craters. Metabolism stalls, as the body goes into a semi-hibernating funk triggered by the winter-like darkness. Energy supplies get stored as fat. Underworked, the body turns on itself, inflammaging creeping unchecked through our organs. And the longer this goes on, the harder it becomes for the body to restore balance when it finally does get outside.

Back home in Vermont that spring, my sleep settled into a Hadza-like six and a half hours. Not a ton, but it felt qualitatively different, deeper and more restorative. I wasn't fighting the day anymore.

Adjustments, however, were necessary. For one thing, work called. MacBook Air may be the enemy, but it's also the lifeline. The need to put in full days made my morning time all the more crucial. I rose early, pushing outside in the frosty dawn to catch whatever light I could. But the spring New England lightscapes couldn't have been more different from the brown hundred-mile views of Arizona. My morning rambles took me along mossy forest floors dripping with ferns and wildflowers, the ground a web of dappled patterns, what the Japanese call *komorebi*, "sunlight leaking through trees."

My path led me along an old stone wall at the forest edge, a hillside of new grass falling away below. A family of deer grazed across it, happy for the fresh salad after a winter of dry twigs. As the plants around me captured the red, blue, and ultraviolet photons they needed, they sprayed a shower of unwanted greens and infrareds in all directions.

I thought back to my zen afternoon in the Sleep Lab. If green light can melt pain and agitation and nurture a contemplative state, and if infrared can improve metabolism and cognition, is that why woodlands feel so good to us? Japan's forest bathers seem to have figured this out long ago. Our early ancestors, sheltering under acacia trees to contemplate the green meadows beyond, never knew anything else. Perhaps we still need that garden a little bit.

All I know is that it felt as if I'd found some sort of sweet spot. I weaved between the dappled shade of the forest and the strong light of the meadows, working the edges of the day. It felt like the photon equivalent of a colorful meze platter. With light that nice, LEDs truly seemed the stuff of asylums and nightmares. I stepped from the shadows into the bright field and continued my course. The deer bolted, bouncing like springboks. The sun warmed my skin in the cool morning air. Whatever malaise had gripped me last winter was gone. I'd found the ultimate solarium.

CHAPTER 15

RIGHT WITH LIGHT

Putting Heliotherapy into Practice

By now it should be clear that there are lots of reasons to be sun curious. The benefits I look for from sunlight fall into six areas:

- Lower blood pressure and better cardiovascular health
- Improved mood and reduced stress
- Healthier immune function and reduced chronic inflammation
- Increased metabolism and mitochondrial function
- Better sleep
- Enhanced energy, alertness, and cognition

I don't tend to think about these things separately. The body is a nest of interlocking systems. Everything affects everything else. Everything happens at once. If you increase your daily circadian amplitude, for example, with bright days and dark nights, you are going to sleep better, and that is going to impact mood, cognition, energy, immune function, cardio health, and so on. So it makes sense to think of all the heliotherapy perks as one package.

That's why, compared to any other wellness routine, the tenets of heliotherapy are ridiculously simple. I actually laid out the whole program in the introduction to this book. It's seven words long:

Get sun. Not too much. Go outside.

That's it. There's no special equipment to buy. No supplements to take. No app to tell you what to do. All you really need is your own common sense. The rule, the experts tell me, is to make sure you never burn. Sunburns are a symptom of severe skin damage. There's nothing healthy about them. And because it's so easy to head for the beach with every intention not to burn and still come back extracrispy, my personal rule is *never come anywhere close to burning.*

That rule works for everyone, but in practice it means different things for different people because we all have different skin tones. What burns me might not burn you, and what lowers my blood pressure six points might barely budge yours. That's one of the biggest considerations in designing your heliotherapy program, and it's where mainstream recommendations for sun exposure tend to fall short.

SKIN IN THE GAME

One of the biggest shortcomings of past sun recommendations is that they have failed to take skin type into account; the guidelines were only appropriate for people with the fairest skin and assumed everyone else could just follow along without harm. Such approaches are finally getting scrapped, and they should be. "A one-size-fits-all approach isn't productive when it comes to sun exposure recommendations," the University of Texas dermatologist Adewole Adamson told me. "It can cause harm to some populations."

The Penn State anthropologist Nina Jablonski concurred, telling me, "We're going to improve overall public health much more by relaxing this guideline than by enforcing it." Everyone knows this needs to happen, but it will still be accompanied by plenty of grumbling. "The problem is that many dermatologists feel it's taken so long to get people to use sunscreen, so they're extremely hesitant to go public with recommendations about relaxing sunscreen use," Jablonski said. "They're afraid that people will just throw away their bottle of sunscreen, throw away their hats, and cast aspersions on the profession for having asked them to do this stuff in the first place."

But we will get there, Jablonski said. "I think it will be important to adjust these things and essentially make a prescription for personalized sun exposure and diet, depending on where you are, who you are, and under what conditions you're living. It'll just be part of the personalized medicine packages that we will have in future generations. . . . I think this will be done, but it's going to be done in dribs and drabs over the course of the next decade."

Jablonski told me that a few years ago, and sure enough, the dribs and drabs are finally coming. Surprisingly, one of the most notable revisions came in Australia, skin-cancer capital of the world. In 2023, a consortium of representatives from Australia's top health organizations, including MS Australia, Cancer Council Australia, Skin Cancer College Australasia, and the Australasian College of Dermatologists, came together in a two-day summit to hash it out. Their new forty-four-page "Position Statement" (which is available on the Australian Skin and Skin Cancer Research Centre website) includes new official recommendations for sun exposure. "Completely avoiding sun exposure is not optimal for health," it reads.

The project was led by the University of Queensland skin-cancer expert Rachel Neale, one of Australia's most prominent voices on sun protection. "We decided it was time to come up with a new message that actually recognized that the balance of the harms and benefits of sun exposure is not the same for everybody," she told me. Neale said the move was inspired by new evidence of the benefits of sun exposure, along with surprisingly high levels of vitamin D deficiency in Australia and signs that people with darker skin were not getting the right information about the risks and benefits to them, in part because doctors themselves were confused about the science.

The new guidelines are directed toward health-care providers as well as the public. The recommendations are tailored to different times of year, situations, and skin tones. People with very fair skin are counseled to use sun protection at all times except when the UV index is less than 3, in which case "sun protection is not generally advised." People with intermediate skin are advised to spend "sufficient time outdoors with ample skin exposed to avoid vitamin D deficiency." For people with dark skin, "benefits of sun exposure outweigh the harms." Adults in this group "do not need to routinely protect the skin from the sun."

The recommendations get extremely granular, triangulating between locations, months, UV index, and skin tone. Neale emphasized to me that it's better to enlist as much skin as possible in your "daily dose of sunshine," so you're not putting undue burden on any one area. "Regular short exposures with as much skin exposure as possible is better for vitamin D production than trying to use a small amount of skin for longer and doing a whole heap of damage to that bit," she said. In other words, ten minutes of naked cartwheels in the backyard might not be a terrible idea.

Overall, the recommendations are still quite conservative, but they stand as the most enlightened official recommendations to date—which some experts find ironic. Adewole Adamson, for example, told me it seems absurd that sunny Australia, with an extremely white population, has "a better official position than do US organizations where nonwhite Americans will outnumber white Americans in the next twenty years."

Rachel Neale also stressed to me that the new recommendations were specifically tailored to sunny Australia and wouldn't make sense for countries at high latitudes, such as Northern Europe. "The messaging for those countries needs to be really different. Certainly there doesn't need to be this same push to be putting on sunscreen every day. . . . For those countries, I think the messaging about 'just don't get burnt' is probably not a bad one."

SIZING UP SUNSCREEN

This is a book about sunlight, not sunscreen, but weighing the risks and benefits of sun exposure inevitably leads to questions about the pros and cons of sunscreen. It's amazing those questions aren't asked more often. Sunscreen is a drug—just ask the FDA—and like any drug, it should only be used when it's going to improve your health. Do not be tricked into thinking that you need to wear it 365 days a year, indoors or out. I'll again quote the sunscreen expert Brian Diffey: "This is totally the wrong advice. . . . It's not a good habit. It's a bad habit. You shouldn't do it."

That's also the assessment of Robyn Lucas: "Absolutely use sun protection to avoid skin cancer. But sunscreen should be the last thing. Sunscreen is for when you can't do anything else. Most other times, clothing is a better choice."

The recommendation to go hog wild with sunscreen is based on the assumption that there is no downside to doing so, but we know that's not accurate. Sunscreens clearly aren't terribly bad for you—millions of people have used them for decades, and so far no one is sprouting an extra head—but experts believe they carry a small amount of risk.

You can include the FDA among the concerned. It has flagged the most common active chemicals in sunscreens as potential hormone disruptors and has refused to certify them as safe, pending more testing.

For many years, we were told this wasn't a problem because the chemicals in sunscreens weren't absorbed through the skin. Then, in 2008, the CDC published a study showing that 97 percent of Americans were contaminated with oxybenzone, the most common and concerning of the sunscreen chemicals, and that sunscreen was overwhelmingly responsible. It turns out oxybenzone and other related chemical filters penetrate the skin with ease. Oops.

By 2019, the FDA's concern had grown, in part because the new advice to apply sunscreen generously every day meant that people were being exposed to far more of these chemicals than before. The agency asked the sunscreen makers to supply better safety-testing data on oxybenzone and its kin, but none of them complied, so the following year the FDA did its own studies and found that the seven most concerning ingredients were absorbed into the bloodstream in large quantities. A single application of sunscreen resulted in concentrations of the chemicals in urine, blood, and breast milk that were beyond the allowable threshold, and they were still present in the body weeks later.

That result finally woke the media from its coma and generated a wave of alarming stories. "It is outrageous that over a decade ago, nearly every American tested by CDC had oxybenzone in their blood, an ingredient linked to hormone disruption, and still manufacturers have resisted safety testing," David Andrews, the senior scientist who assesses sunscreens for the Environmental Working Group, told CNN at the time.

In 2021, mounting evidence that the sunscreen running off snorkelers' bodies could also damage coral reefs led Hawaii to ban sunscreens containing oxybenzone and octinoxate, two of the most worrisome chemicals. Since then, many other common chemical filters have been implicated. According to the National Oceanic and Atmospheric Association (NOAA), these chemicals may cause mutations, deformities, and birth defects in coral, fish, shellfish, and other marine life. As of 2026, places that have banned reef-damaging sunscreens include Hawaii, Key West, Palau, Aruba, Bonaire, the US Virgin Islands, and marine parks in Thailand and Mexico. In addition, many other countries strongly encourage adoption of reef-safe sunscreens but have not enacted formal bans.

We also know that many skin-care products, including sunscreens, can be contaminated with phthalates and PFAS, two extremely worrisome classes of chemicals, and that these chemicals easily migrate into the body. Phthalates have been connected to neurological disorders, reproductive abnormalities, and metabolic dysfunction. A 2024 study at George Mason University found elevated levels in children's urine that reflected their use of skin-care products.

Just as concerning is the presence of PFAS, the notorious "forever chemicals" believed to cause cancer, liver damage, and other health problems. Even products that don't list PFAS in their ingredients are often contaminated with them because they can leach out of the plastic containers. Once again, toxicologists found that these forever chemicals are absorbed through the skin at shockingly high rates. And whether they wind up in you or the local waters, true to their name, they never go away. "Any chemist over the last five decades that looked at those chemicals could have told you that they're never gonna break apart in the environment, that they'll end up potentially accumulating in people," David Andrews told me. "And yet we have a broken system in place that has enabled the use and release of incredible quantities of these chemicals in everything from the food supply to cosmetics and sunscreens. Let's not do this again."

Let's not. It's worth reiterating that no one thinks the risks from sunscreen chemicals are severe, and everyone agrees they are valuable tools. But there's growing concern among experts that the directive to automatically apply sunscreen every day is not in anyone's best interests. "We need to use sunscreens wisely so that we gain the health benefit from their use while at the same time limiting possible harm to ourselves and the environment," says Brian Diffey.

Diffey has been calling for restraint for nearly thirty years, starting with his 1998 commentary in the *British Journal of Dermatology*, "Sun Protection: Have We Gone Too Far?" He answered his own question in the affirmative in the *BJD* in 2002, reporting, "Virtually no benefit is gained from using UV protective products from October to March in the U.K." He later expanded that assessment to other places, concluding that below thirty degrees latitude (Mexico and the southern United States) sunscreen should be used year-round, at forty degrees (mid-Atlantic states and Southern Europe) it can be skipped in winter, and above fifty degrees (Canada and Northern Europe) it's only necessary during the summer months.

So once again, this seems to be a situation where common sense should overrule public inanity. I don't hesitate to use sunscreen when I need it, which means anytime I might burn without it. One big point in sunscreen's favor is that it blocks UV but not the visual and infrared wavelengths, so you are going to get many of sunlight's energizing and healing benefits without any risks. It also allows you to be in the sun safely, where you reap the circadian advantages mediated through the eyes. In addition, sunscreen does let a small percentage of UV through, so it might actually be a useful tool for titrating your dosage.

On the other hand, in 2025, Australian investigators, led by Rachel Neale, released a major study showing that daily use of SPF 50 significantly impedes the production of vitamin D. By the end of the yearlong study, nearly half the participants who used the SPF 50 were vitamin D deficient, and that's in sunny Australia. Likely this is true for other important sun-associated molecules as well, including those that seem to boost cognitive function and tune the immune system.

For all the reasons above, I avoid anything higher than SPF 30.

BUT WHAT IF YOU HATE SUNSCREEN?

Fortunately, there are lots of good goop-free strategies for controlling your sun dose. The most obvious is simply limiting your time in the sun. But the different physical characteristics of different wavelengths of light open up a number of other strategies, depending on which benefits you seek:

- **Clothing.** Remember that infrared light is the most abundant part of the solar spectrum, and it passes right through normal clothing. So anytime you are out and about in a hat and long-sleeve shirt, you are absorbing lots of infrared and probably making your mitochondria happy, as well as maintaining good circadian alignment through your eyes. You do have to be out and about, though.

- **Time of day.** Because the atmosphere filters out shorter wavelengths more than longer ones, morning and evening have little ultraviolet light, especially UVB, the most damaging. Morning is also the best time to get your circadian machinery cranking.

- **Shade.** As discussed, plants absorb red, blue, and ultraviolet light but they bounce green and infrared. So anytime you are shielded from direct sunlight but surrounded by green leaves, you are bathing in both green and infrared photons. The combo is going to facilitate a soothing, meditative state, boost mitochondrial function, heal tissue, and reduce inflammation. The shade of a tree on a summer day can still have dozens of times as much light as an indoor office, which is what makes it such a sweet spot.

FOUR RULES

Given all these considerations, my heliotherapy plan tends to follow four general principles:

- **Get out every day.** In summer, I try to spend the last few hours of the day outside doing something active—in sunlight if it's not too intense, or in the shade if it is. Bare skin when possible, but clothing or shade if there is any chance of burning. In winter, I like to get out in the middle of the day, soaking up whatever sunlight I can get. (In a low-latitude environment, this isn't so necessary. You'll get your sun just going about your day.) In general, when life allows, the more bright light in my day, the better.

- **Morning magic.** I consider morning light exposure nonnegotiable. It supercharges mental and physical performance with zero downside. If I can squeeze in an hour's walk in the crisp morning air, that's even better. Light on skin is a big bonus, but not possible in all seasons in many places.

- **Let the light in.** I have to spend much of the day working in my office, which is not ideal from a circadian perspective, but that's life. I just try to make sure my working space gets a lot of natural light. If you can't do that, make sure the space is brightly lit with artificial light—not as good, but a lot better than dim lighting.

- **Go dark.** In the evening, I try to keep the light level fairly low as bedtime approaches. I haven't found it necessary to use red light bulbs or any of the more extreme measures, but I have switched to the "warmer" LED bulbs, which produce much less blue light. I've found that if I read on a tablet in a dark room, I can bank down the display incredibly low and there's no negative impact on my sleep. When it is time for sleep, I like a bedroom so dark you can barely navigate.

NO CHILD LEFT INSIDE

In this book I've spent a lot of time exploring how one's sun needs can change depending on skin tone, location, and season. But what about seasons of life? We could use a lot more evidence in this area, but the science that exists suggests that this should be a key part of the discussion.

Kids are much more sensitive to sunlight than adults, and we know melanoma is closely associated with sunburns during childhood. Kids' small bodies and highly absorbent skin are also much more susceptible to the chemicals in sunscreen. Experts recommend using clothing instead of sunscreen on kids as much as possible, reserving sunscreen for spots likely to get burnt.

At the same time, kids who stay inside through childhood have higher rates of many diseases later in life, as well as worse cognitive performance, so the focus should be on getting them outside and playing, whatever it takes. I like Robyn Lucas's approach: "What is it going to hurt for your kids to get outside for an hour a day? You're not asking for much, and you've got the potential for huge benefits!" That doesn't mean parents need to be basting their kids on the beach like rotisserie chickens. "We're not asking them to go out and get sunburned. We're just asking them to run around outdoors for thirty minutes." Shade is great, just get them away from their devices and out of the house.

But in giving kids a daily dose of sunshine, the best time of all might be before they are born. As I've mentioned, kids whose moms get more sun exposure during pregnancy have lower rates of certain autoimmune diseases. And a study of 422,000 Scottish schoolchildren found that the kids whose moms got the most sun exposure during pregnancy had just half the rate of learning disabilities compared to kids of the moms who got the least, and there was a perfect dose-dependent relationship through five different levels of exposure. Similar relationships have been seen for other neurodevelopmental conditions, all in northern countries.

All this strongly suggests that pregnant women should avoid sun deprivation.

SUN AFTER SIXTY?

"I don't believe in aging," Virginia Woolf wrote, shortly after turning fifty. "I believe in forever altering one's aspect to the sun." She meant it as a philosophy, not a personal wellness program, but there are reasons to believe that, as one ages, one should literally alter one's aspect to the sun.

Consider the list of health problems associated with aging. Metabolism slows. Blood pressure rises and insulin resistance grows. Chronic inflammation creeps in. Cognition declines. Sleep gets spotty. Depression sets in. These are all conditions improved by natural light, and the benefits are immediate. Our ability to synthesize vitamin D and other compounds also declines with age, so we need more time in the sun to get the same production.

In contrast, the risks of sun exposure seem to decline as we age. Remember the large studies showing that burns early in life are strongly associated with melanoma, while sun exposure later in life has little association. To be clear, getting burned is always a bad idea, but there's no sign that nonburning sun exposure late in life has any real downside.

So it looks like we need a Sun After Sixty movement. By the time someone has reached their seventh or eighth decade, the risk/benefit ratio is tilting hard in the direction of sunlight. It might be time to loosen up a little bit. Yet many senior citizens get checked into retirement centers and never get out again. Imagine the payoff to public health if we let them back into the garden.

YOU ARE ENERGY

As a last thought, I want to circle back to where I started a couple of hundred pages ago. Lots of practical takeaways can be gleaned from the information in this book—concrete actions and practices that can improve health, performance, and mindset. But also a new richness comes from thinking in terms of energy, not atoms. We are both things at once—the difference is simply in the eye of the beholder—but the future of medicine clearly lies in tracing the flow of electrons through our mitochondria and fixing it when it goes off course. When you see yourself as a standing wave of energy, shaped by atoms but not defined by them, you make better decisions about health and longevity. You think differently about food, movement, sound, light, the very star around which you circle. I think about this every time I catch the sunrise (which I do quite a lot these days), and it has altered my aspect to the coming day in deeply satisfying ways. Where once I might have worried about how to protect myself from all the celestial dangers I'd heard so much about, I now look forward to the extraordinary gifts coming my way once again.

ACKNOWLEDGMENTS

Writing a book that questions the conventional wisdom means you get a lot of noes from people and institutions who aren't comfortable with such lines of inquiry. That makes the yeses from those who are intrigued all the more appreciated. Some of the research for this book was done while I was a Media Fellow at the Nova Institute for Health, which proved invaluable. (Shout-out to Bud Brainard, the Yoda of light.) The Nova Institute is curating an extraordinary community of scholars and thinkers who are forging a new understanding of what it takes to have healthy people on a healthy planet, and I was happy to be a part of it.

I was also happy to be a Knight Science Journalism Fellow. That year at MIT allowed me to unplug from all the propaganda regarding sunlight and health and go back to first principles. All scientific detective stories should work that way.

Some of the other research for this book was done while I was reporting a handful of magazine stories. I'm deeply grateful to my wonderfully inquisitive editors and their publications, in which some of this information first appeared in different form: Elizabeth Hightower Allen at *Outside*, Kiera Butler at *Mother Jones*, Daniel Engber at *The Atlantic*, Josh Fischman at *Scientific American*, and Matthew Sherrill at *Harper's*.

And then there are the scientists themselves, without whom none of this information would exist. A big shout-out to the thousands of researchers whose studies and papers compose the foundation of this book, and especially the ones gracious enough to take the time to explain their work to me in direct interviews: Mohamed Abou-Elwafa Abdallah, Adewole Adamson, David Andrews, Bruce Armstrong, Rob Bilott, George Brainard, Ad Brand, Frank de Gruijl, Chris Dibben, Ulysse Dormoy, Martin Feelisch, Bob Fosbury, Cedric Garland, Stuart Harrad, Prue Hart, Lauren Hartstein, Nina Jablonski, Glen Jeffery, Kat Kennedy, Margaret Kripke, Annette Langer-Gould, David Leffell, Greg Leonard, Henry Lim, Pelle Lindqvist, Robyn Lucas, John MacMahon, Rachel Neale, Stefan Pilz, Oddný Ragnarsdóttir, Andrzej Slominski, Corey Washington, Richard Weller, Antony Young, Kathy Young, and Scott Zimmerman.

I met many of those scientists at conferences organized by the Sunshine Health Foundation, the late Allen Miller's extraordinary endeavor to jump-start scientific conversations about sunshine and health. It's safe to say that the conversation would be years behind without Allen's work, and he is sorely missed.

Huge thanks to Deacon and Debbie Drawdy for sharing their parched piece of paradise with me in Arizona, and for getting the water working.

On the words front, I got great help from Angela Miller, as always. Thanks to Adrienne Anderson for the Tick Tock tip and to Kim "like playing tetherball with fire" Madalinski for feeding me her best lines. It was thrilling to find myself back with editor extraordinaire Kathy Belden for yet another book, and to collaborate with the talented team at Scribner. And through it all, even in the depths of winter, Mary Elder Jacobsen kept the home light shining. *Post tenebras lux.*

GLOSSARY

ALAN: Artificial light at night. This refers to all nonnatural illumination. More than 80 percent of the world's population is subject to ALAN, which is believed to be detrimental to sleep and health.

amplitude: The difference between the high point and the low point in anything that oscillates, such as a wave (in which it would be the difference between the trough and the crest). A greater amplitude in circadian cycles (brighter days and darker nights) is associated with improved sleep, metabolism, and longevity.

antioxidant: Any compound that protects against the damage caused by free radicals, which are molecules that have been oxidized (i.e., have had their electrons snatched by oxygen). Free radicals are unstable and can attack other molecules in cells, damaging them. Antioxidants bond with free radicals, neutralizing them. Breathing, producing energy, sun exposure, and many other parts of daily life generate free radicals, which must be neutralized each day with antioxidants. The body gets some antioxidants from food (mostly fruits and vegetables) and also produces its own. Some of the most powerful antioxidants are vitamin D, melatonin, and melanin.

autoimmune disease: Any disease in which the immune system attacks the body's own cells. Examples include Crohn's disease, psoriasis, type 1 diabetes, rheumatoid arthritis, and multiple sclerosis.

basal cell carcinoma (BCC): The most common form of skin cancer, accounting for 80 percent of all cases, and the least dangerous. BCCs almost never become invasive, but they do need to be removed, which can cause scarring.

blue blockers: Glasses designed to block the passage of blue photons while admitting other colors. They are used to reduce exposure to blue light in the evenings (from LED bulbs and screens), which can suppress melatonin production and affect sleep.

chronobiology: The study of periodic cycles in living things. Mostly daily or yearly ones.

circadian: The twenty-four-hour cycle of the earth and biology's adaptations to it.

color: A range of the light spectrum that triggers particular receptors in the eye. There is no color in nature, just an unbroken spectrum of light with different frequencies and wavelengths. We have different receptors tuned to certain frequencies, and when these receptors absorb light with that frequency, they send a signal to the brain that registers as a color. Because different colors of photons have different amounts of energy and different abilities to penetrate the body, they have different impacts on health.

confounder: Any type of variable that connects behaviors among subjects in observational studies in ways that can distort results. For example, people who exercise tend to be healthier, and they are also less likely to smoke, so a study of the health effects of smoking must take into account exercise as a confounding variable. (A study needs to compare smokers who exercise against smokers who don't; otherwise, it might unintentionally be comparing the health effects of exercise instead.) Classic confounding variables that are "adjusted for" in all quality studies include smoking, exercise, diet, education, income, gender, and age.

cortisol: A hormone produced by the adrenal glands that improves the body's ability to tackle challenges by increasing alertness and energy release. It makes our cells run at higher speed. Cortisol is good in the short term, but too much stress can lead to chronically high cortisol levels and health problems. Sun exposure encourages a healthy release of cortisol during the day, and darkness causes levels to drop at night—in the absence of stress.

dermis: The thick layer of skin, below the epidermis, that gives skin its strength and flexibility. It includes sweat and oil glands and blood vessels. Red and infrared photons make it to the dermis, but blue and ultraviolet ones don't make it past the epidermis.

EEG: Electroencephalogram. A device used to measure brain waves and mental states.

epidemiology: The study of patterns of disease in populations, used to identify causative factors. Many of the first observations that people exposed to more sunlight were healthier came from epidemiologists.

epidermis: The thin, outermost layer of skin that interacts directly with the environment. It contains the common skin cells that protect the body and that also communicate with the brain, immune system, and endocrine system through chemical messengers. It also contains the melanocytes, which make melanin. Skin cancers start in the epidermis. It is exposed to all wavelengths of light.

Fitzpatrick scale: A classification system developed by the dermatologist Thomas Fitzpatrick in the 1970s that divides human skin tones into six categories, based on amount of melanin and response to ultraviolet light. A type I is very fair and easily burns; a type III has olive skin and tans readily; and a type VI has very dark skin and never burns, even under strong sunlight.

free radical: A molecule that has become electrically charged and seeks to acquire an electron from a nearby molecule to return to a balanced state, destabilizing that molecule in turn. Free radicals can cause major damage to cells, but they are generated by all the tasks of daily life, and under normal circumstances the body easily neutralizes them using antioxidants.

frequency: The rate at which a system (whether a light wave, molecule, or tuning fork) oscillates between a high point and a low point. The frequency of a photon determines which molecules in the body can absorb it.

Hadza: A hunter-gatherer group in Tanzania's Great Rift Valley that still lives a traditional lifestyle without artificial light or industrial foods. The Hadza have been extensively studied for insights into their excellent health and the deleterious effect of modern lifestyles on ours.

heliotherapy: The use of sunlight to treat disease. Originally applied to treating rickets and tuberculosis, now being explored for many more systemic conditions.

homeostasis: The stable internal environment (in terms of temperature, blood pressure, oxygenation, etc.) necessary for a body to function well. More broadly, the systems the body employs to stay in this happy zone despite all the curveballs the world throws at it.

hormesis: The phenomenon in which small challenges elicit an adaptive response from the body that strengthens it in the long run. Experts increasingly believe hormesis is a major factor in health and longevity. Exercise is a form of hormesis, causing minor damage that forces the body to mobilize its repair mechanisms and increase its future capacity. So is intermittent fasting. Sensible sun exposure is another form of hormesis, resulting in a stronger, more capable physiology.

hormone: A chemical messenger produced by one part of the body that communicates with other parts of the body via the bloodstream or lymphatic system (as opposed to the nervous system).

immunosuppression: The phenomenon of suppressing the immune system's response to a pathogen or a stressor. Can be bad if it allows an infection or cancer to proliferate unchecked or can be good if it prevents the rejection of a skin graft or organ transplant or quells an autoimmune disease. In the 1970s, ultraviolet light was discovered to be immunosuppressive.

inflammaging: A type of low-grade, chronic inflammation that tends to increase with age among most people in the industrial world, but is rarely seen in hunter-gatherer societies living traditional lifestyles. It is believed to be a major factor in many noncommunicable diseases, including cancer, cardiovascular disease, and cognitive dysfunction.

inflammation: The initial series of actions taken by the immune system in response to a new injury, infection, or other aberration. It includes an increase of blood flow to the area and a mobilization of defensive cells to destroy the problem. It frequently causes damage, but after the problem is resolved, it should be followed by an anti-inflammatory phase in which the damage is healed and the tissue returned to normal. Inflammation is a healthy and vital part of daily function, but it can become a problem if it is too frequent or never resolves, leading to chronic disease.

infrared: The spectrum of light with wavelengths longer than that of red light. It accounts for 53 percent of the solar radiation that reaches us. We can't see infrared light because we don't have detectors for it in our eyes, but it penetrates deeply into the body and is believed to have a number of physiological effects we are only beginning to understand. It passes right through clothing, so we are absorbing it anytime we are outside. But it doesn't pass through buildings, so we are deprived of it whenever we are indoors.

laser: A device that emits a concentrated beam of photons of light having the same wavelength (color) and direction of travel. Used worldwide for photobiomodulation (red-light therapy).

latitude gradient: A phenomenon first noted in the early twentieth century in which many diseases become more frequent the farther people live from the equator. It is now widely believed to be linked to sun deprivation.

LED: Light-emitting diode. A type of illumination that produces light by exciting electrons in a semiconductor material, causing them to release photons as they jump back and forth between two states of excitement. (When they drop from a high-energy state to a low one, they emit a photon of energy.) All modern lights are LEDs, which are much more efficient than older bulbs because their semiconductors emit a specific wavelength of light. Older incandescent bulbs, by contrast, produced a wide range of photons, mostly infrared, which we couldn't see (though could feel as heat), so much of their electricity was wasted. Most LEDs produce a narrow range of photons centered on the blue part of the spectrum, which means they have an unnatural composition compared to sunlight. They also tend to have a strong suppressive effect on melatonin production.

lesion: A change in tissue, such as that caused by a wound or a skin cancer. Moles are also considered lesions, even though they are usually benign.

lightscape: The local light environment that surrounds you and impacts your physiology and consciousness. Sunlight and artificial light are the biggest factors in a lightscape, but it is also affected by clouds, shade, time of day, colors of surrounding objects, and many other physical details.

lux: A measure of light defined as one lumen per square meter, or a candle flame at arm's length. A romantic dining room might be 100 lux, a bright office 400 lux, a cloudy day outdoors 10,000 lux, and direct sunshine 100,000 lux. People are often shocked to discover how much brighter it is outdoors, even in the shade.

melanin: A pigment made in the skin that absorbs solar radiation, protecting the skin from damage. Melanin is also a powerful antioxidant. People with dark skin have a much higher melanin content than people with pale skin and are much less susceptible to sun damage.

melanocytes: Special cells in the basal layer of the epidermis that produce melanin.

melanoma: A cancer of the melanocytes. Can be aggressive and dangerous. Fortunately, invasive melanoma accounts for just 1 percent of skin cancers. Closely associated with sunburns during childhood in fair-skinned people.

melanoma in situ: The earliest stage of melanoma, also referred to as stage 0 melanoma. Believed to be wildly overdiagnosed. Most of the rise in melanoma incidence in the past half century is due to the increase in stage 0 melanomas, and most of these are probably not actual melanomas, but are removed anyway to err on the side of caution.

melatonin: A hormone produced in the brain by the pineal gland in response to darkness. As melatonin suffuses the brain in the evening, cells go quiet and sleep comes. Light exposure, especially blue wavelengths, arrests melatonin production and can interfere with sleep. Melatonin is also an extremely capable antioxidant that is produced inside all of our cells and is used daily to clean up free radicals. This melatonin doesn't leave the cells and isn't related to sleep.

Mendelian randomization: A technique for assessing the effect of some environmental factor in which natural variations in individuals' ability to produce some compound are measured against results. For example, people who possess certain gene variants are naturally better at producing vitamin D. If those people are healthier overall than other people in the same region who are exposed to the same amount of sunlight, then vitamin D must be good for you. (Spoiler alert: they're not.) Mendelian randomization was one of the techniques that helped establish the surprisingly minimal impact on health of vitamin D and pointed to other factors in sunlight that were responsible for the benefits.

mercury-vapor lamp: An early form of lighting that produced light by running an electrical current through a glass of vaporized mercury. The lamps produce a lot of ultraviolet light. This is normally filtered by the glass, but the lamps can be modified to deliver UV instead. Such lamps have been used in research and were notoriously used as sunlamps in the early twentieth century before UV's role in skin cancer was widely understood.

metabolism: The endless chemical reactions that occur in a living body that allow it to produce energy and the compounds it needs to survive. Essentially, the molecular pathways of biological transformation. The more efficient your metabolism, the healthier you are.

microbiome: The community of microbiotic organisms that inhabit a particular ecological niche. The microbiome in our gut has a significant impact on our health, and so does the microbiome that lives on our skin. The makeup of that community can change drastically depending on what it is exposed to—an area of health we are just starting to explore. It's now believed that a flourishing, sun-adapted skin microbiome can serve as a first line of defense against sun damage and certain diseases.

mitochondria: Tiny, bacteria-like entities that inhabit our cells and produce most of our energy, among many other vital tasks. The ancestors of mitochondria were independent bacteria. Complex, multicellular life (such as us) began when one of these bacteria merged with a larger single-celled organism and they began to divvy up tasks. We have thousands of mitochondria inside most of our cells, and a great deal of our health and longevity depends on keeping them in great shape. They seem to like infrared light a lot.

multiple sclerosis (MS): An autoimmune disease in which the immune system attacks the myelin sheaths that coat the nerves, leading to a variety of breakdowns in electrical signaling. This can result in loss of motor function, speech, vision, cognition, and many other problems. MS is sometimes called the sunlight disease because exposure to sunlight during childhood correlates to much-lower incidence later in life.

myelin: A pliable, cholesterol-like coating that insulates nerve cells the same way plastic coating insulates electrical wires in buildings. Not only does myelin prevent electricity from leaking out of the nervous system, recent research indicates it may also store energy for the neurons. (A "proton capacitor" if you want to get technical.) Like a battery, the myelin sheath stores an energy gradient that can then be discharged into the energy-hungry neurons, improving their performance. This may help to explain why people who suffer from MS, which damages the myelin, experience a lot of fatigue.

nanostructure: A tiny molecular structure that functions at the quantum level. Plants contain nanostructures in their leaves designed to capture individual photons of light energy. Some scientists now believe we contain nanostructures designed to harvest infrared light energy as well.

narrowband UV: A specific wavelength of ultraviolet light used therapeutically to suppress an overactive immune response without increasing the risk of skin cancer. Typically, narrowband-UV lamps produce ultraviolet light with a wavelength of 311 nanometers, which has not been shown to cause DNA damage at the levels used in clinics. Often used to treat psoriasis and other skin disorders, it is now being explored as a treatment for MS and other systemic autoimmune diseases.

neurotransmitter: Any molecule used by neurons to communicate in the brain and greater nervous system. We now know that the skin also produces a constant stream of neurotransmitters used to coordinate with the brain and influence behavior, mood, and homeostasis.

nitrate: A molecule composed of three oxygen atoms and one nitrogen atom. Nitrates are plentiful in root vegetables and leafy greens. We store them in large quantities in our skin, where they can be converted by sunlight into nitric oxide, a vital compound that regulates vasodilation, blood pressure, and many other bodily processes.

nitric oxide: A signaling molecule composed of one nitrogen atom and one oxygen atom. Nitric oxide is used throughout the body to regulate blood flow, inflammation, bone strength, and many other basic processes. When sunlight hits skin, it liberates nitric oxide and releases it into the bloodstream, resulting in dilation of the blood vessels and lower blood pressure. This is believed to be one of the reasons sun exposure reduces cardiovascular disease.

observational study: A scientific study of data from real-world populations, as opposed to data generated by trials in a clinic. It is harder to control for confounding variables in an observational study because people who have certain things in common often have other behaviors in common as well. On the other hand, observational studies can be the best way to discover new factors that affect health. For instance, the benefits of the Mediterranean diet were discovered by observational studies that compared eating habits and heart disease in different countries.

opioid: A class of compounds that attach to receptors in the brain, block pain signals, and trigger feelings of bliss. The body produces natural opioids as a reward mechanism for behaviors it wants to encourage (sex, eating, sun exposure), but opioid drugs, such as opium, heroin, morphine, and fentanyl, can hack the same pleasure centers. They are very addictive.

photobiomodulation (PBM): A type of therapy, first developed in the 1960s and 1970s, that uses red or infrared light to relieve pain, stimulate wound healing, and reduce inflammation.

photoimmunology: The science of how light affects the immune system.

photon: The fundamental unit of light. A packet of pure energy. Photons are often pictured as particles, and it can be useful to imagine them as tiny balls of light traveling through space, but because they have no mass, they aren't really particles. Sometimes it's helpful to picture light as an electromagnetic wave of energy perturbing space-time, and sometimes it's more helpful to picture it as a stream of colored balls. I do both in this book. Welcome to quantum mechanics.

phototherapy: The use of light as a therapeutic tool, whether artificial (red-light therapy, narrowband-UV lamps, SAD lamps) or natural sunlight.

POMC: Proopiomelanocortin, a large molecule produced by the skin in response to sunlight that is then cleaved by enzymes into three different hormones. One stimulates the production of melanin in the skin, deepening tans; one triggers the production of cortisol, which increases metabolism and activity; and one releases natural opioids in the brain, triggering feelings of well-being.

randomized clinical trial (RCT): The gold standard of scientific studies. RCTs are done under controlled conditions, when every variable can be kept the same except the one being studied. Subjects are randomly sorted so that the groups being studied aren't different in any important way. RCTs are the best way to determine what effect changing one factor might have, but they sometimes lack real-world applicability, since many effects can be multifactorial or chaotic and in the real world conditions are constantly changing.

reactive oxygen species (ROS): A type of free radical caused by energizing molecules containing oxygen. An abundance of ROS can damage cells, but they are also caused by daily life: breathing, exercising, metabolizing food, exposure to light. Under normal circumstances, the body is able to neutralize them with antioxidants.

reverse causation: A potential stumbling block in observational studies, where the perceived cause and effect are actually reversed. For instance, when scientists first noticed that people who got more outside time were healthier and proposed that something about being outdoors was responsible, others argued that it could be reverse causation: maybe people who were already healthier were more likely to go outdoors. (TL;DR: It's both.)

rickets: A disease caused by vitamin D deficiency in which bones don't harden sufficiently, leading to bowed legs, enlarged skulls, curved spines, pain, and weakness. Rickets became prevalent in children during the early phases of the industrial revolution, due to a lack of sun exposure in the factories, tenement buildings, and soot-filled cities of the time.

scatter: The way light changes direction after interacting with molecules. When photons pass through translucent materials, they can ricochet in multiple new directions. This happens in the air (the sky looks blue because blue photons in particular tend to scatter off molecules of the atmosphere) and in the body. This pachinko effect allows some photons to penetrate the body quite deeply.

slow-wave sleep: A vital stage of sleep, distinct from REM, in which brain cells are inactive and no dreaming occurs. This is when the brain cleans itself of the day's waste products and repairs any damage. Slow-wave sleep often suffers in people with sleep problems.

spectroscopy: The study of the light spectrum and how it interacts with matter.

spectrum: The range of colors (and wavelengths) of light associated with a particular phenomenon. The spectrum of "white" sunlight includes all the visible colors, as well as ultraviolet and infrared. The spectrum of a typical "bright white" LED includes mostly blue light, along with some green, yellow, and orange, and no ultraviolet or infrared.

SPF: Sun protection factor. A measure of how much longer one can stay out in the sun without burning, compared to not using the sunscreen. So on a day when you could stay in the sun for thirty minutes before burning, with a SPF 30 sunscreen you could stay for nine hundred minutes, or fifteen hours. But SPFs are problematic. They refer only to UVB, so they aren't a good indication of how well the sunscreen blocks UVA. And recent revelations in Australia have shown that the SPF number listed on the bottle can be woefully inaccurate.

squamous cell carcinoma (SCC): A common form of skin cancer, responsible for about one in every five cases. SCCs, which form in the outermost layers of skin cells, are generally not dangerous, but they are more likely to cause trouble than BCCs. They are also the type of skin cancer most effectively prevented by sunscreen.

tuberculosis: A disease caused by the bacterium *Mycobacterium tuberculosis*. It can affect both the skin and the lungs and was rampant in the nineteenth century, when it was known as the white plague, or consumption, and killed millions of people. The cutaneous form was successfully treated by sunlight, which killed the bacteria. Sanatoriums also treated pulmonary tuberculosis with fresh air and sunlight. This had no direct effect on the bacteria but did bolster the health of the patients through systemic pathways.

ultraviolet (UV): The range of the solar spectrum with wavelengths shorter than violet light. We can't see UV, as we have no receptors in our eyes tuned to it. It is the highest-energy portion of the spectrum, high enough to ionize atoms and break bonds in molecules. That can cause damage to DNA and lead to skin cancer, but it is also used by the body to produce vitamin D and nitric oxide, among other compounds. We divide UV into two sections: UVA and UVB. UVB comprises the highest-energy solar light that reaches Earth and can cause direct DNA damage. UVA is much more abundant but much less energetic. It rarely damages DNA but does generate a lot of free radicals, which can indirectly lead to cancer. In general, UVB is more closely associated with basal cell carcinoma and squamous cell carcinoma, while UVA is more closely associated with melanoma.

urocanic acid: A molecule that is abundant in the skin, making up about 1 percent of the dry weight of the epidermis, and which plays multiple roles in health and disease. It functions as a natural moisturizer and sunscreen, as well as an excellent food source for beneficial microflora on the skin, but it also causes immunosuppression when hit by UV and converted from its trans to its cis isomer. Most intriguingly, after being exposed to sunlight, it travels from the skin to the brain, where it metabolizes into glutamate and enhances the activity of neurons, resulting in improved learning, memory, and mood.

UV index: A scale introduced in the 1990s to help people gauge how intense the sun would be on that day. A UV index below 3 is mild, 3–5 is moderate, 6–7 is high, 8–10 is very high, and 11 or higher is scary. UV indexes are published daily as part of the weather forecast for most places on Earth. Note that the number for the day applies only to the peak hours around midday. It can be quite a bit lower in the morning and late afternoon, and it's never a problem at sunrise or sunset.

vitamin D: A hormone produced in the skin with the help of UV light, which breaks a bond in a cholesterol molecule that allows it to change its shape and become vitamin D. It should never have been labeled a vitamin, since vitamins are supposed to be compounds that the body needs but can't make itself. The more sun exposure you get, the more vitamin D you make. It's used in the skin as an antioxidant and anticarcinogen, and it is used throughout the body to regulate calcium and keep bones strong, and it may even be an essential regulator of circadian and seasonal genes, but its previous reputation as a miracle drug has not panned out in clinical trials. It turned out other things about sunlight were responsible for most of the benefits credited to vitamin D.

wavelength: The distance in any type of wave between one crest and the next. With the light waves that reach us, ultraviolet has the shortest wavelengths (and the most energy) and infrared has the longest (and least energy), with the visible wavelengths in the middle in their usual rainbow order. Light's wavelength dictates which molecules can absorb it and which will deflect it.

RESOURCES

BOOKS

***American Sunshine*, by Daniel Freund**
History of America's relationship with the sun from the tuberculosis era through the first part of the twentieth century.

***The Big Fat Surprise*, by Nina Teicholz**
A thorough takedown of nutritionism and how it got diet and health so wrong. A master class in dissecting the fallibility of scientific institutions.

***Eating the Sun*, by Oliver Morton**
Challenging but amazing book on the science of photosynthesis.

***Exercised*, by Daniel Lieberman**
The best book yet written on exercise. New insights on why we dread it, why we still need to do it, and why it is beneficial in unexpected ways.

***Health and Light*, by John Ott**

Ott pioneered the art of time-lapse photography in the 1930s and went on to produce some classic films for Walt Disney in which he made flowers "dance" by manipulating their light environments between photos. That work led to a fascination with the ways that organisms responded to changes in light, and alarm in what our unnatural light environments were doing to us. He founded the Environmental Health and Light Research Institute, wrote *Health and Light* in 1973, and produced the film *Exploring the Spectrum* in 1974. Ott's ideas were never taken seriously by the scientific establishment, and some of his ideas were outlandish, but his thinking was decades ahead of his time.

***Heliotherapy*, by Auguste Rollier**

The 1923 original, now quite outdated but still full of prescient insights about the relationship between the body and sunlight.

***Rise and Shine*, by Simon Carter**

A history of changing attitudes toward sunlight in the UK.

***The Science of Sleep*, by Wallace Mendelson**

In-depth explanation of what we know about the biochemistry of sleep, heavy on the graphics.

***Skin: A Natural History*, by Nina Jablonski**

Definitive work on the science and semiology of skin, by the anthropologist who has looked at it the closest.

***Sleepless*, by Annabel Abbs-Streets**

One woman's account of her insomnia and how it changed her attitude toward night.

***Sun Seekers*, by Lyra Kilston**

Seminal work that connects tuberculosis and the desire to escape the grimy industrial cities of the nineteenth century to the German body-culture movement, modernist architecture, and the migration to California in the early twentieth century in search of a new way of living.

***Suntanning in 20th Century America*, by Kerry Segrave**

A fun romp through the history of sunscreen and sun attitudes, with lots of laughs along the way.

***The Vitamin D Solution*, by Michael Holick**

The definitive guide to vitamin D, by the leading expert on the subject.

***What the Body Knows*, by John Trowsdale**

Excellent explanation of how the immune system works.

***Why We Sleep*, by Matthew Walker**

Great guide to how sleep works (and doesn't) and how to improve your own.

LIGHTS

Allay

Green light for migraine relief, designed by Rami Burstein, the Harvard neurologist who first discovered the effect.

Cytokind

Leading producer of narrowband UV lamps for psoriasis, vitiligo, eczema, MS, and other autoimmune conditions. The website includes a wealth of information on the science behind the lamps.

Solius

This Puget Sound–based start-up is the first to bring to market a next-generation, over-the-counter, FDA-cleared light for raising vitamin D levels. The nifty panel is extremely portable and delivers a precisely focused beam of 293nm UVB light. Just a minute a day significantly raises D levels with zero safety risk. The interface is through your phone. The team behind it is extremely knowledgeable. Expect to see a lot of these around.

PODCASTS

Huberman Lab

Among A-list podcasters, Andrew Huberman is the one who has delved into the real science on light and health. His explanations of the body's circadian mechanics have been especially good. Many of the scientists in this book have been guests on the show.

MedCram

The podcast is good but the YouTube channel is excellent, with fun graphics and animations illustrating the fundamental principles of health, drilling all the way down to the level of mitochondria and electron transport. Hosted by the affable Loma Linda physician Roger Seheult, it's an excellent primer on photons and many other things.

Regenerative Health

The Australian MD Max Gulhane is a leader in the holistic health movement, focusing on ancestral diets and lifestyles and regenerative farming. His podcast episode list is a who's who of experts on sunlight and circadian health, and he's quite good at explaining the basics to a nonscientific audience.

Ricci Flow Nutrition

Another Australian, Cameron Borg, specializes in cutting-edge explorations of how light and life interact on the quantum level, and how health is built from these fundamental flows of energy. (Ricci flow is a mathematical concept explaining how natural flows of energy can sculpt an irregularly shaped object or process back into symmetry and balance.) Mind-blowing interviews with some of the most interesting scientists and nutritionists on the planet. His Substack feed is equally provocative.

SUN EXPOSURE RECOMMENDATIONS

Australian Skin and Skin Cancer Research Centre

assc.org.au

Look for their position statement "Balancing the Harms and Benefits of Sun Exposure."

SUNSCREENS

Environmental Working Group

ewg.org/sunscreen/

Since 2007, EWG has been publishing its annual evaluation of sunscreens on the US market. Of the more than two thousand products in its latest roundup, only 23 percent were deemed safe and acceptable. Those are the only ones you should use. The site includes a wealth of information on what's in sunscreens and what you need to worry about.

ORGANIZATIONS

Daylight Academy
daylight.academy
European association of architects, scientists, and lighting professionals dedicated to spreading the word about the importance of natural daylight to human health. In addition to funding research, they organize workshops and conferences and lectures and host an annual Daylight Symposium.

The Guy Foundation
theguyfoundation.org
Founded by UK pharma pioneer Geoffrey Guy in 2018, the Guy Foundation is a quantum-biology think tank and funder of cutting-edge research on the interface of life and thermodynamics, with its own dedicated laboratories and extensive outreach programs. Its excellent speaker series, including the 2025 Autumn Series on Light, is available for free on the website.

Sunshine Health Foundation
sunshinehealthfoundation.org
Founded by sunlight advocate Allen Miller, the Sunshine Health Foundation funds research and scientific conferences on the health impacts of light. The website is an indispensable clearinghouse for scientific information on sun exposure.

NOTES

This is a fairly exhaustive list of the books, papers, and other sources of information for the facts cited in this book. It also includes some additional thoughts and anecdotes from me that were too tangential or scientific for the main text but might appeal to my fellow light nerds. To save space, I include only the core details needed to look up a source: first author, title, publication, date. Certain journals pop up again and again, and for those I just use their initials. They are *BJD* (*British Journal of Dermatology*), *BMJ* (*British Medical Journal*), *JAAD* (*Journal of the American Academy of Dermatology*), *JAMA* (*Journal of the American Medical Association*), *JID* (*Journal of Investigative Dermatology*), *PNAS* (*Proceedings of the National Academy of Sciences*), *PPS* (*Photochemical & Photobiological Sciences*). I would love to have included a thousand more intriguing papers, so if anything hooks you and you want to go deep, let's talk.

INTRODUCTION

2 *In the mirror, I looked a little bit puffy:* One possible factor in this was discovered in 2017. Cutaneous fat cells can detect blue light, and they use that as a signal to decide how much fat to store. When they don't detect blue light (winter, clothes), they store more fat and slow metabolism. See Ondrusova, "Subcutaneous White Adipocytes Express a Light Sensitive Signaling Pathway Mediated via a Melanopsin/TRPC Channel Axis," *Scientific Reports*, 2017.

2 *a team of dermatologists at Harvard Medical School confirmed:* Fell, "Skin Beta-Endorphin Mediates Addiction to UV Light," *Cell*, 2014.

2 *POMC is actually three essential compounds linked together:* For some good overviews on POMC (and other endocrine compounds generated by the skin), see Slominski, "How UV Light Touches the Brain and Endocrine System Through the Skin, and Why," *Endocrinology*, 2018; and Slominski, "Photo-Neuro-Immuno-Endocrinology: How the Ultraviolet Radiation Regulates the Body, Brain, and Immune System," *PNAS*, 2024.

3 *I actually think of cortisol as the "in the zone" hormone:* Cortisol also improves mood. See Hoyt, "Positive Upshots of Cortisol in Everyday Life," *Emotion*, 2016. I also highly recommend Daniel Lieberman's book *Exercised* (Vintage Books, 2020) for a full account of the myriad ways exercise improves health by initially stressing the body.

5 *a paradigm-shifting study appeared in* Cell*:* See Zhu, "Moderate UV Exposure Enhances Learning and Memory by Promoting a Novel Glutamate Biosynthetic Pathway in the Brain," *Cell*, 2018; Lowe, "Sunlight and the Brain," *Science*, 2018; and Chantranupong, "Sunlight Brightens Memory and Learning," *Cell*, 2018. The researchers used mice (which are crepuscular) in their experiments, so the belief is that the effects would be even stronger in a diurnal species such as us.

5 *And not just mice:* Komulainen, "Long-term Residential Sunlight Exposure Associated with Cognitive Function Among Adults Residing in Finland," *Scientific Reports*, 2022.

6 *The science consistently showed that people:* For a few good overviews on the science documenting sunlight's health benefits, see Riedmann, "Beneficial Health Effects of Ultraviolet Radiation," *PPS*, 2025; Alfredsson, "Insufficient Sun Exposure Has Become a Real Public Health Problem," *International Journal of Environmental Research and Public Health*, 2020; Weller, "Sunlight: Time for a Rethink?," *JID*, 2024; Hart, "Exposure to Ultraviolet Radiation in the Modulation of Human Diseases," *Annual Review of Pathology*, 2019; and Byrne, "How Much Sunlight Is Enough?," *PPS*, 2014.

6 *Parkinson's, myopia, respiratory infections:* For Parkinson's, a meta-analysis found that sun exposure reduced the risk, but vitamin D supplementation had no effect. See Zhou, "The Association Between Vitamin D Status, Vitamin D Supplementation, Sunlight Exposure, and Parkinson's Disease: A Systematic Review and Meta-Analysis," *Medical Science Monitor*, 2019. Another UK Biobank study found that the risk of Parkinson's was significantly lower among those who spent more time outside: Hu, "Association of Time Spent Outdoors with the Risk of Parkinson's Disease: A Prospective Cohort Study of 329,359 participants," *BMC Neurology*, 2024.

6 *one massive UK study attached light-sensing watches:* Windred, "Brighter Nights and Darker Days Predict Higher Mortality Risk," *PNAS*, 2024.

10 *fifteen scientists with expertise in these fields recently published:* Alfredsson, "Insufficient Sun Exposure."

10 *A decade ago, most experts thought they knew:* For an excellent overview of D's changing fortunes, see Aschwanden, "The Rise and Fall of Vitamin D," *Scientific American*, 2024.

11 *Top publications such as* The New England Journal of Medicine*:* Cummings, "VITAL Findings—a Decisive Verdict on Vitamin D Supplementation," *New England Journal of Medicine*, 2022.

11 *Top magazines such as* Scientific American *ran features debunking it:* Aschwanden, "Rise and Fall of Vitamin D."

13 *"This is totally the wrong advice," the dermatologist:* Baird-Murray, "Do You Actually Need to Wear SPF Every Day? Separating Fact from Fiction," *Vogue*, 2025.

14 *"you're walking out of my office":* One of the studies Weller is referencing in this statement was done in Denmark in 2013. It looked at the entire Danish population above age forty (more than 4 million people) and found that those with skin cancer were less likely to get heart attacks or strokes and less likely to die from any cause. The study adjusted for age, economics, and physical activity. Brøndum-Jacobsen, "Skin Cancer as a Marker of Sun Exposure," *International Journal of Epidemiology*, 2013.

19 *the multitude of benefits people receive when they're outside:* Twohig-Bennett, "The Health Benefits of the Great Outdoors: A Systematic Review and Meta-Analysis of Greenspace Exposure and Health Outcomes," *Environmental Research*, 2018.

CHAPTER 1 **SOL MATES**

27 *this skin microbiome forms a kind of living sunscreen:* For more on the protective qualities of the skin microbiome, see Souak, "Challenging Cosmetic Innovation: The Skin Microbiota and Probiotics Protect the Skin from UV-Induced Damage," *Microorganisms*, 2021; Burns, "Ultraviolet Radiation, Both UVA and UVB, Influences the Composition of the Skin Microbiome," *Experimental Dermatology*, 2019; Nakatsuji, "A Commensal Strain of Staphylococcus Epidermidis Protects Against Skin Neoplasia," *Science Advances*, 2018; and L'Orphelin, "The Skin Microbiome: A New Key Player in Melanoma, from Onset to Metastatic Stage," *Pigment Cell and Melanoma Research*, 2024.

27 *The skin of lifeguards:* Harel, "Skin Microbiome Bacteria Enriched Following Long Sun Exposure Can Reduce Oxidative Damage," *Research in Microbiology*, 2023.

28 *In lower-light environments, even when people were spending:* One fascinating bit of research in this area was done by Pelle Lindqvist, the Swedish researcher who will figure prominently in later chapters. He found that in low-light environments in Sweden, redheads had better overall survival than other Caucasian Swedes with darker skin and hair. It's clear that the redheaded/freckled/fair-skinned phenotype, with its very different form of melanin, is for some reason advantageous in such environments. See Lindqvist, "Women with Fair Phenotypes Seem to Confer a Survival Advantage in a Low UV Milieu: A Nested Matched Case Control Study," *PLOS One*, 2020.

29 *a team of dermatologists at the University of Manchester:* Felton, "Concurrent Beneficial (Vitamin D Production) and Hazardous (Cutaneous DNA Damage) Impact of Repeated Low-Level Summer Sunlight Exposures," *BJD*, 2016.

30 *an accompanying editorial by Michael Holick:* Holick, "Can You Have Your Cake and Eat It Too? The Sunlight D-lema," *BJD*, 2016.

31 *As the dermatologist Richard Weller:* Weller, "Does Incident Solar Ultraviolet Radiation Lower Blood Pressure?" *Journal of the American Heart Association*, 2020.

CHAPTER 2 CARE OF THE BODY

32 *On a dazzling December weekend in 1929:* The story of Philip Lovell and the Health House has been told many times, including in Lyra Kilston's excellent book *Sun Seekers* (Atelier Éditions, 2019).

33 *"A flat, open roof can be of immense value":* Lovell, Care of the Body, *Los Angeles Times*, 1924.

33 *"For years I have periodically written":* It's worth looking up this column: Lovell, Care of the Body, *Los Angeles Times*, 1929.

35 *Twentieth-century doctors and architects:* For historical information on ancient sun attitudes and heliotherapy in the nineteenth and twentieth centuries, see Jarrett, "A Short History of Phototherapy, Vitamin D and Skin Disease," *PPS*, 2016; Albert, "The Evolution of Current Medical and Popular Attitudes Toward Ultraviolet Light Exposure," *JAAD*, 2002; Randle, "Suntanning: Differences in Perceptions Throughout History," *Mayo Clinic Proceedings*, 1997; and the following books: Rollier, *Heliotherapy* (Oxford Medical Publications, 1923); Ott, *Health and Light* (Pocket, 1973); Segrave, *Suntanning*

in 20th Century America (McFarland, 2005); Carter, *Rise and Shine* (Berg, 2007); Freund, *American Sunshine* (University of Chicago Press, 2012); and Kilston, *Sun Seekers.*

35 *This question was first asked by Theobald Palm:* Palm, "The Geographical Distribution and Aetiology of Rickets," *Practitioner*, 1890.

36 *Not until 1913 would the mystery begin to be solved:* There are many histories of rickets available online, including Jones, "100 Years of Vitamin D: Historical Aspects of Vitamin D," *Endocrine Connections*, 2022; Chesney, "Theobald Palm and His Remarkable Observation: How the Sunshine Vitamin Came to Be Recognized," *Nutrients*, 2012; and Wheeler, "A Brief History of Nutritional Rickets," *Frontiers in Endocrinology*, 2019. See also Michael Holick's book *The Vitamin D Solution.*

37 *a consensus emerged: vitamin D was produced:* Jones, "100 Years of Vitamin D."

38 Heliotherapy. *It's a nice word:* For histories of heliotherapy, see Jarrett, "Short History of Phototherapy"; Albert, "Evolution of Current Medical"; Rollier, *Heliotherapy*; Carter, *Rise and Shine*; and Freund, *American Sunshine.*

38 *By the late nineteenth century it was responsible:* Ritchie, "Once a Leading Killer, Tuberculosis Is Now Rare in Rich Countries— Here's How It Happened," *OurWorldinData*, 2025.

39 *The father of heliotherapy was a sickly doctor:* For information on Finsen and his institute, see the Nobel Prize Committee's biography on its website. See also Grzybowski, "From Patient to Discoverer—Niels Ryberg Finsen (1860–1904)—the Founder of Phototherapy in Dermatology," *Clinics in Dermatology*, 2012.

39 *"I long for it," he later wrote:* Randle, "Suntanning."

40 *they did get many things right:* Quotes are from Rollier's 1923 book, *Heliotherapy.*

41 *in 2025 the dermatologist Richard Weller found:* See Weller's 2025 lecture at the Guy Foundation, available at the foundation's website: Weller, "The Health Effects of Sunlight, UV, and Blue Light," Guy Foundation, 2025.

42 *By then characters were firming up all over Europe:* Information on Europe's heliotherapy wave can be found in Saleeby, "The Advance of Heliotherapy," *Nature*, 1922; in Saleeby's 1923 book, *Sunlight and Health*; and in Lyra Kilston's book *Sun Seekers.*

42 *Solaria dotted the American landscape, too:* See Freund, *American Sunshine*; and Kilston, *Sun Seekers.*

43 *"excluding ourselves from the light:* Loignon, "Bringing Light to the World: John Harvey Kellogg and Transatlantic Light Therapy," *Journal of Transatlantic Studies*, 2022.

44 *"The climate is so healthy":* Praslow, "The State of California: A Medico-Geographical Account," *Newbegin*, 1939. Quoted in Lyra Kilston's book *Sun Seekers.*

44 *Chandler came under the sway of the naturopath Philip Lovell:* For information on Lovell, see his Care of the Body columns in the *Los Angeles Times*; Lyra Kilston's book *Sun Seekers*; and the charming account of his great-nephew Gary Marmorstein, "Steel and Slurry: Dr. Philip M. Lovell, Architectural Patron," *Southern California Quarterly*, 2002.

45 *vita-glass, a new product designed to let ultraviolet rays:* See Freund, *American Sunshine.*

46 *a 1916 article on advances in war surgery:* "Modern Science and War Surgery," *Scientific American*, 1916.

46 *Schoolhouses were redesigned with high banks of windows:* Gyure, "The Transformation of the Schoolhouse: American Secondary School Architecture and Educational Reform, 1880–1920" (University of Virginia dissertation, 2001).

46 *"The treatment of disease by sunlight":* Saleeby, "The Advance of Heliotherapy," *Nature*, 1922.

47 *"Light is life, and lack of it causes ill health":* "Sunlight Therapy's Value Certain in Tuberculosis," *New York Times*, 1924.

47 *"the antirachitic effects of exposure to sunlight":* "Sunshine and Health," *JAMA*, 1925.

47 *The US government distributed a pamphlet:* Freund, *American Sunshine.*

47 *In 1926,* The Lancet *called sunbathing:* Worthington, "Sun-Bathing in Upper Egypt," *Lancet*, 1926.

47 *Cornell was one of many institutions to establish:* Albert, "Evolution of Current Medical."

48 *the president of the Chicago Board of Health wrote:* Bundesen, "Sunshine and Health," *Ladies' Home Journal*, 1938.

48 *No country took that exhortation more to heart than Germany:* For information on German nudism and body culture, see Freund, *American Sunshine*; Kilston, *Sun Seekers*; and Toepfler, *Empire of Ecstasy: Nudity and Movement in German Body Culture, 1910–1935* (University of California Press, 1997).

48 *That paled in comparison to Moscow:* See Chase, "Confessions of a Sun-Worshiper," *Nation*, 1929.

48 *a community of nudists who had commandeered:* "When Nude Sunbathing Caused a Riot in Hendon," *Londonist*, 2022.

48 The Lancet *weighed in with its support:* "Naked and Unashamed," *Lancet*, 1932.

49 *Stuart Chase knew he was preaching to the choir:* Chase, "Confessions of a Sun-Worshiper."

49 *Wheatena felt it necessary to point out on its box:* The Wheatena ad was in *Ladies' Home Journal* in 1929. See Albert, "Evolution of Current Medical."

50 *as* Harper's Bazaar *observed in 1900:* Murray, "The Summer Girl's Complexion," *Harper's Bazaar*, 1900.

50 *The beaches turned into "public rotisseries":* "Beauty and Sunburn," *New York Times*, 1931; and Adams, "Modern Worshipers of That Old God, the Sun," *New York Times*, 1929.

50 Collier's *dubbed it "ultraviolet insanity":* Seinfel, "The Burning Question," *Collier's*, 1933.

51 *"A golden tan is the index of chic!":* Randle, "Suntanning."

51 *Stores filled with UV machines:* See Albert, "Evolution of Current Medical"; and Segrave, *Suntanning in 20th Century.*

51 *The wildly popular lamps had the blessing:* Albert, "Evolution of Current Medical."

51 *"In the future . . . we shall sleep in beds":* "Says We Will Sleep Under Violet Rays," *New York Times*, 1931.

51 *Schools gave up on natural lighting:* Freund, *American Sunshine.*

CHAPTER 3 SLIP, SLOP, SLAP

53 *The German dermatologist Paul Gerson Unna:* Unna, *The Histopathology of the Diseases of the Skin* (William F. Clay, 1896).

53 *the Chicago dermatologist James Nevins Hyde:* Hyde, "On the Influence of Light in Production of Cancer of the Skin," *American Journal of the Medical Sciences*, 1906.

54 *even* The Journal of the American Medical Association *responded:* "Ultraviolet and Cancer," *JAMA*, 1929.

54 *Just the year before, England's George Marshall Findlay:* Findlay, "Ultra-Violet Light and Skin Cancer," *Lancet*, 1928.

54 *"Every physician knows that farmers and seamen":* "Sunlight and Cancer," *New York Times*, 1933.

55 *"The evidence is such as to constitute a warning":* "Smoker's Cancer Laid to Sunburn," *New York Times*, 1941.

55 *The 1940s also saw the rise of a new product:* For information on the early days of sunscreen, see Freund, *American Sunshine* (University of Chicago Press, 2012); and Segrave, *Suntanning in 20th Century America* (McFarland, 2005).

55 *The first effective sunscreen was developed in 1942:* Straight from the horse's mouth: Luckiesh, "Protective Skin Coatings for the Prevention of Sunburn," *JAMA*, 1946.

56 *an airman named Benjamin Green:* See Segrave, *Suntanning in 20th Century*; and "Sunscreen: A History," *New York Times*, 2010.

57 *three hundred light researchers decided unanimously:* Diffey, "The UVB Content of 'UVA Fluorescent Lamps' and Its Erythemal Effectiveness in Human Skin," *Physics in Medicine and Biology*, 1983.

59 *But what really kicked the business:* Segrave, *Suntanning in 20th Century*.

59 *the goal was to make sunscreen something:* Belkin, "A Fierce Fight over the Sun," *New York Times*, 1986.

59 *"There is not enough ultraviolet light on Earth":* Sloan, "Suncare Marketers Raise Their Screens," *Advertising Age*, 1987.

60 *the FDA gave in to the inevitable:* Sweet, "Healthy Tan—a Fast-Fading Myth," *FDA Consumer*, 1989.

60 *the "Australia sunscreen scandal":* Turnbull, "Australia Sunscreen Scandal Grows as More Products Pulled off Shelves," BBC, 2025.

61 *"It turns out to be hard to show":* "A Spectrum of Sun Protection," *Consumer Reports*, 1998.

61 *a meta-analysis of the twenty-nine studies:* Da Silva, "Use of Sunscreen and Risk of Melanoma and Non-Melanoma Skin Cancer: A Systematic Review and Meta-Analysis," *European Journal of Dermatology*, 2018.

61 *an exposé in* Mother Jones *magazine:* Castleman, "Beach Bummer," *Mother Jones*, 1993. Also see Castleman's 1998 follow-up: Castleman, "Sunscam," *Mother Jones*, 1998.

62 *"There is no evidence that sunscreens protect you":* "Sunscreen: The Burning Facts," Environmental Protection Agency, 2006. For other overviews of the current evidence, see Brunner, "Malignant Melanoma: The Relationship Between Sunscreen Use and Cancer Risk—a Systematic Review and Meta-Analysis," *Anticancer Research*, 2025; Arisi, "Sun Exposure and Melanoma, Certainties and Weaknesses of the Present Knowledge," *Frontiers in Medicine*, 2018; and Huncharak, "Topical Sunscreens and the Risk of Malignant Melanoma: A Meta-Analysis of 9,067 Patients from 11 Case–Control Studies," *American Journal of Public Health*, 2002.

62 *In 2009, the UK sunscreen expert Brian Diffey:* Diffey, "Sunscreens—Expectation and Realization," *Photodermatology, Photoimmunology and Photomedicine*, 2009.

62 *"There is no high-quality experimental evidence":* Raimondi, "Melanoma Epidemiology and Sun Exposure," *Acta Dermato-Venereologica*, 2020.

62 *A 2023 study by dermatologists from Harvard:* Lapides, "Possible Explanations for Rising Melanoma Rates Despite Increased Sunscreen Use over the Past Several Decades," *Cancers*, 2023.

CHAPTER 4 THE MYSTERIES OF MELANOMA

64 *Melanoma is a different beast:* Or possibly several beasts. The different behaviors of different types of melanoma mean that we are probably lumping several different diseases with different etiologies under one umbrella term. See Yeh, "Melanoma Pathology: New Approaches and Classification," *BJD*, 2021.

64 *while melanoma isn't closely associated with lifetime:* See Wu, "History of Severe Sunburn and Risk of Skin Cancer Among Women and Men in 2 Prospective Cohort Studies," *American Journal of Epidemiology*, 2016; and Savoye, "Patterns of Ultraviolet Radiation Exposure and Skin Cancer Risk: The E3N-SunExp Study," *Journal of Epidemiology*, 2017. But sunburns aside—and perhaps counterintuitively, perhaps not—more time spent outside during childhood is protective against melanoma. See Kaskel, "Outdoor Activities in Childhood: A Protective Factor for Cutaneous Melanoma? Results of a Case-Control Study in 271 Matched Pairs," *BJD*, 2001.

65 *Harvard dermatologist David Fisher has shown:* Mitra, "An Ultraviolet-Radiation-Independent Pathway to Melanoma Carcinogenesis in the Red Hair/Fair Skin Background," *Nature*, 2012.

65 *That idea was confirmed by a 2025 study:* Whiteman, "Validity of Risk Prediction Tool for Invasive Melanoma," *JAMA Dermatology*, 2025.

65 *a 2005* Journal of the National Cancer Institute *study:* Berwick, "Sun Exposure and Mortality from Melanoma," *Journal of the National Cancer Institute*, 2005.

65 *A French study of sixty-seven thousand women:* Mahamat-Saleh, "Mediterranean Dietary Pattern and Skin Cancer Risk: A Prospective Cohort Study in French Women," *American Journal of Clinical Nutrition*, 2019.

65 *a 2024 presentation by the UK skin-cancer expert Amaya Viros:* Viros, "Ultraviolet Light-Induced Collagen Degradation Inhibits Melanoma Invasion," *Nature Communications*, 2021.

66 *Melanoma was clearly very different:* Over the years, numerous papers have tried to make sense of this. See Planta, "Sunscreen and Melanoma: Is Our Prevention Message Correct?," *Journal of the American Board of Family Medicine*, 2011; Huncharak, "Topical Sunscreens and the Risk of Malignant Melanoma: A Meta-Analysis of 9,067 Patients from 11 Case–Control Studies," *American Journal of Public Health*, 2002; Godar, "Increased UVA Exposures and Decreased Cutaneous Vitamin D(3) Levels May Be Responsible for

the Increasing Incidence of Melanoma," *Medical Hypotheses*, 2009; Godar, "Cutaneous Malignant Melanoma Incidences Analyzed Worldwide by Sex, Age, and Skin Type over Personal Ultraviolet-B Dose Shows No Role for Sunburn but Implies One for Vitamin D3," *Dermato-Endocrinology*, 2017; Moan, "Epidemiological Support for an Hypothesis for Melanoma Induction Indicating a Role for UVA Radiation," *PPS*, 1999; Gorham, "Do Sunscreens Increase Risk of Melanoma in Populations Residing at Higher Latitudes?," *Annals of Epidemiology*, 2007; and Garland, "Rising Trends in Melanoma: An Hypothesis Concerning Sunscreen Effectiveness," *Annals of Epidemiology*, 2007.

66 *"Both UVA and UVB are carcinogenic":* Wallis, "Bring Back the Parasol," *Time*, 1983.

67 *For every photon of UVB that reaches:* Estimates vary. See Bernerd, "The Damaging Effects of Long UVA (UVA1) Rays: A Major Challenge to Preserve Skin Health and Integrity," *International Journal of Molecular Science*, 2022.

67 *At a 1990 conference, they argued:* Angier, "Theory Hints Sunscreens Raise Melanoma Risks," *New York Times*, 1990.

67 *But the Garlands pushed their case in a 1992 letter:* Garland, "Could Sunscreens Increase Melanoma Risk?," *American Journal of Public Health*, 1992.

68 *the meta-analysis in the* European Journal of Dermatology*:* Da Silva, "Use of Sunscreen and Risk of Melanoma and Non-Melanoma Skin Cancer: A Systematic Review and Meta-Analysis," *European Journal of Dermatology*, 2018.

68 Mother Jones *did its exposé:* Castleman, "Beach Bummer," *Mother Jones*, 1993. Also see Castleman's 1998 follow-up: Castleman, "Sunscam," *Mother Jones*, 1998.

69 *In her paper in the* Journal of the National Cancer Institute*:* Wolf, "Effect of Sunscreens on UV Radiation-Induced Enhancement of Melanoma Growth in Mice," *Journal of the National Cancer Institute*, 1994.

69 *Kripke's study drew coverage in* The New York Times*:* Kolata, "Mouse Study Raises Doubts on Sunscreens," *New York Times*, 1994.

69 *a triumphant letter from the editor in chief:* Klein, "Sunscreen May Be Skin Cancer Villain," *New York Times*, 1994.

69 *Kripke's experiment "should cause red flags":* Davis, "Boggy Ground," *Drug & Cosmetic Industry*, 1994.

70 *the terrible pattern in that era:* The UK sunscreen expert Brian Diffey is one of many who have made this point, arguing in 2005 that the old sunscreens probably didn't prevent melanoma but the new ones should: Diffey, "Sunscreens and Melanoma: The Future Looks Bright," *Photobiology*, 2005.

Further support for this idea came from a 2008 study in the *Journal of the American Academy of Dermatology* showing that new, broad-spectrum sunscreens prevented the classic UV-induced immunosuppression believed to be a big factor in melanoma, but old-school sunscreens that blocked only UVB did not: Moyal, "Broad-Spectrum Sunscreens Provide Better Protection from Solar Ultraviolet-Simulated Radiation and Natural Sunlight-Induced Immunosuppression in Human Beings," *JAAD*, 2008.

70 *In 2010, the industry finally got the study:* Green, "Reduced Melanoma after Regular Sunscreen Use: Randomized Trial Follow-Up," *Journal of Clinical Oncology*, 2011.

70 *what it actually compared is exemplary use:* One of the papers that has pointed this out is Planta, "Sunscreen and Melanoma."

70 *For other skin cancers:* Green, "Daily Sunscreen Application and Betacarotene Supplementation in Prevention of Basal-Cell and Squamous-Cell Carcinomas of the Skin: A Randomised Controlled Trial," *Lancet*, 1999; and van der Pols, "Prolonged Prevention of Squamous Cell Carcinoma of the Skin by Regular Sunscreen Use," *Cancer Epidemiology, Biomarkers, and Prevention*, 2006.

71 *according to the* Journal of the American Academy of Dermatology*:* Weyers, "A. Bernard Ackerman—the 'Legend' Turns 70," *JAMA*, 2006.

71 *"The field is just replete with nonsense":* Kolata, "I Beg to Differ; a Dermatologist Who's Not Afraid to Sit on the Beach," *New York Times*, 2004.

72 *when they sequence the DNA of normal, functional:* Yizhak, "RNA Sequence Analysis Reveals Macroscopic Somatic Clonal Expansion Across Normal Tissues," *Science*, 2019.

72 *One of the dermatologists who has looked:* Here are Adamson's key papers: Welch, "The Rapid Rise in Cutaneous Melanoma Diagnoses," *New England Journal of Medicine*, 2021; Lopes, "UV Exposure and the Risk of Cutaneous Melanoma in Skin of Color," *JAMA Dermatology*, 2020; Marchetti, "Melanoma and Racial Health Disparities in Black Individuals—Facts, Fallacies, and Fixes," *JAMA Dermatology*, 2021; and Adamson, "Ecological Study Estimating Melanoma Overdiagnosis in the USA Using the Lifetime Risk Method," *BMJ Evidence-Based Medicine*, 2024.

73 *In one study, pathologists were asked:* Frangos, "Increased Diagnosis of Thin Superficial Spreading Melanomas: A 20-year Study," *JAAD*, 2012.

74 *He estimated that more than 85 percent:* Adamson, "Ecological Study Estimating."

74 *Ha-ha, responded the Academy:* Kulkarni, "To Improve Melanoma Outcomes, Focus on Risk Stratification, Not Overdiagnosis," *JAMA Dermatology*, 2022.

74 *In 2007 doctors called out the American Cancer Society:* Aschwanden, "Doctors Balk at Cancer Ad, Citing Lack of Evidence," *New York Times*, 2007.

75 *Adewole Adamson estimates that true melanoma rates:* Adamson, "Ecological Study Estimating."

76 *that's exactly what we do see:* Paulson, "Age Specific Incidence of Melanoma in the United States," *JAMA Dermatology*, 2020.

76 *which are linked to the development of melanoma:* Nurla, "Recent-Onset Melanoma and the Implications of the Excessive Use of Tanning Devices—Case Report and Review of the Literature," *Medicina*, 2024.

76 *tanning beds have been shown to induce far more mutations*: Gerami, "Molecular Effects of Indoor Tanning," *Science Advances*, 2025.

77 *It caused no damage at all:* Felton, "Concurrent Beneficial (Vitamin D Production) and Hazardous (Cutaneous DNA Damage) Impact of Repeated Low-Level Summer Sunlight Exposures," *BJD*, 2016.

77 *In a second experiment:* Shih, "Fractional Sunburn Threshold UVR Doses Generate Equivalent Vitamin D and DNA Damage in Skin Types I–VI but with Epidermal DNA Damage Gradient Correlated to Skin Darkness," *JID*, 2018.

78 *Dark skin is also much better at repairing:* Brenner, "The Protective Role of Melanin Against UV Damage in Human Skin," *Photochemistry and Photobiology*, 2008.

78 *In the United States, skin cancer is seventy times:* Brenner, "Protective Role of Melanin"; and Kaidbey, "Photoprotection by Melanin—a Comparison of Black and Caucasian Skin," *JAAD*, 1979.

79 *he wrote in an opinion piece:* Adamson, "In Rare Occasions, Dark-Skinned People Can Get Skin Cancer. But Sunscreens Won't Help," *Washington Post*, 2019.

79 *Adamson backed that up with a 2020:* Lopes, "UV Exposure."

80 *a 2018 Italian study:* Italians who took sunny holidays before or after their melanoma diagnosis also tended to have less severe melanomas. See Gandini, "Sunny Holidays Before and After Melanoma Diagnosis Are Respectively Associated with Lower Breslow Thickness and Lower Relapse Rates in Italy," *PLOS One*, 2013.

81 *But a 2023 study at the University of Pittsburgh:* Smith, "Melanoma Detection in Alaska Native, American Indian, Asian, Black, Hispanic, and Pacific Islander Patients in a Large Skin Cancer Screening Initiative," *JAMA Dermatology*, 2023.

81 *"This is an almost unfathomable number":* "Screening Won't Solve Racial Disparities in Melanoma Outcomes, Study Suggests," University of Pittsburgh Medical Center, 2023.

CHAPTER 5 CHANGES IN LATITUDE

87 *From 1958 to 1964, Keys and his team:* Keys, *Seven Countries: A Multivariate Analysis of Death and Coronary Heart Disease* (Harvard University Press, 1980). Or just look up the Seven Countries Study website.

88 *Keys's single-minded focus on saturated fat:* The best coverage of the study's shortcomings is Nina Teicholz's book, *The Big Fat Surprise* (Simon and Schuster, 2014).

88 *The other glaring factor that somehow eluded Keys:* But others didn't miss it. David Grimes noted the obvious math that every molecule of dehydrocholesterol converted into D was one less molecule available to be converted into cholesterol: Grimes, "Sunlight, Cholesterol and Coronary Heart Disease," *QJM*, 1996. This ability to lower cholesterol supplies by converting it into vitamin D was later put to the test and confirmed in India: Patwardhan, "Randomized Control Trial Assessing Impact of Increased Sunlight Exposure Versus Vitamin D Supplementation on Lipid Profile in Indian Vitamin D Deficient Men," *Indian Journal of Endocrinology and Metabolism*, 2017. That study aside, most observers think the overall impact on cholesterol levels due to vitamin D conversion is minimal, as less than 5 percent of the dehydrocholesterol in the skin ever gets converted to D. See: Holick, "The D-lemma: Narrow-band UV Type B Radiation Versus Vitamin D Supplementation Versus Sunlight for Cardiovascular and Immune Health," *American Journal of Clinical Nutrition*, 2017.

88 *Keys fell hard for the Mediterranean sunshine:* Keys's enthusiasm for the Mediterranean climate can be found in his book *How to Eat Well and Stay Well the Mediterranean Way* (Doubleday, 1975). For a review of his work, see Nina Teicholz's book, *The Big Fat Surprise.*

89 *a paper in 1932 in the* American Journal of Public Health: Emerson, "Sunlight and Health," *American Journal of Public Health*, 1932.

90 *That dichotomy was pointed out in* The Lancet *in 1936:* Peller, "Carcinogenesis as a Means of Reducing Cancer Mortality," *Lancet*, 1936.

90 *Frank Apperly came up with a better explanation:* Apperly, "The Relation of Solar Radiation to Cancer Mortality in North America," *Cancer Research*, 1941.

90 *the gradient for cancer mortality is highly significant:* There are two sources for the data on cancer mortality and the latitude gradient, both unpublished as this book went to press. The company Cytokind compared SEER data on cancer mortality in health-care facilities in the United States against UV index data for their locations. And the dermatologist Richard Weller has

analyzed UK Biobank cancer data and latitude for a forthcoming paper. He discusses his findings in his 2025 lecture at the Guy Foundation, available at the foundation's website: Weller, "The Health Effects of Sunlight, UV, and Blue Light," Guy Foundation, 2025. See also a 2016 paper that found a strong association between sun exposure and incidence of invasive cancer, but not of mortality: Fleischer, "Solar Radiation and the Incidence and Mortality of Leading Invasive Cancers in the United States," *Dermatoendocrinology*, 2016.

91 *a hugely influential paper in the* Journal of Epidemiology*:* Garland, "Do Sunlight and Vitamin D Reduce the Likelihood of Colon Cancer?," *International Journal of Epidemiology*, 1980.

93 *"Imagine a treatment that could build bones":* Parker-Pope, "Vitamin D, Miracle Drug: Is It Science, or Just Talk?," *New York Times*, 2010.

93 *"More and more studies are revealing the benefits":* Kalish, "Knowing Your Vitamin D Levels Might Save Your Life," *Oprah*, 2010.

95 *One of the first big tests for vitamin D:* Bouillon, "The Health Effects of Vitamin D Supplementation: Evidence from Human Studies," *Nature Reviews Endocrinology*, 2022.

95 *In 2014,* The Lancet *published a review:* Autier, "Vitamin D Status and Ill Health: A Systematic Review," *Lancet Diabetes and Endocrinology*, 2014.

96 *as the prestigious* New England Journal of Medicine *explained:* Cummings, "VITAL Findings—a Decisive Verdict on Vitamin D Supplementation," *New England Journal of Medicine*, 2022.

97 *a UK team looked at the genomes:* Meng, "Phenome-Wide Mendelian-Randomization Study of Genetically Determined Vitamin D on Multiple Health Outcomes Using the UK Biobank Study," *International Journal of Epidemiology*, 2019.

97 *D didn't matter. It was a red herring:* Another fascinating piece of the vitamin D story was too complex to get in the main text of this book, but could turn out to be a big deal. Over the past two decades, a scientist at the University of Alabama named Andrzej Slominski has documented nearly twenty different vitamin D–like molecules produced in the skin by sunlight, an entire family of previously overlooked signaling hormones that have different antioxidant, anti-inflammatory, antimicrobial, and anticarcinogenic properties. These "cutaneous messengers of the sun," as Slominski calls them, are produced by different combinations of light and enzymes in the skin, and while they are all related to vitamin D, they likely each have their own functions and quite possibly work best as a package. And you can't get them from a pill. See Slominski, "Photo-Neuro-Immuno-Endocrinology: How the Ultraviolet

Radiation Regulates the Body, Brain, and Immune System," *PNAS*, 2024; Slominski, "How UV Light Touches the Brain and Endocrine System Through Skin, and Why," *Endocrinology*, 2018; and Slominski, "Biological Effects of CYP11A1-Derived Vitamin D and Lumisterol Metabolites in the Skin," *JID*, 2024.

CHAPTER 6 THE SWEDISH PARADOX

100 *As Lindqvist wrote in the paper:* See Lindqvist, "Thrombotic Risk During Pregnancy: A Population Study," *Obstetrics and Gynecology*, 1999; and Lindqvist, "The Relationship Between Lifestyle Factors and Venous Thromboembolism Among Women: A Report from the MISS Study," *British Journal of Haematology*, 2008.

102 *Sun-seeking Swedish women had dramatically lower:* Lindqvist, "Does an Active Sun Exposure Habit Lower the Risk of Venous Thrombotic Events? A D-lightful Hypothesis," *Journal of Thrombosis and Haemostasis*, 2009.

102 *What about diabetes?:* Lindqvist, "Are Active Sun Exposure Habits Related to Lowering Risk of Type 2 Diabetes Mellitus in Women: A Prospective Cohort Study?," *Diabetes Research and Clinical Practice*, 2010. Also see Lindqvist's follow-up, which established the dose-dependent relationship: Lindqvist, "Sun Exposure and Type 2 Diabetes Mellitus: A Prospective Follow-Up Cohort Study from Southern Sweden," *Anticancer Research*, 2025.

103 *Over the twenty years that the women:* Lindqvist, "Avoidance of Sun Exposure Is a Risk Factor for All-Cause Mortality: Results from the Melanoma in Southern Sweden Cohort," *Journal of Internal Medicine*, 2014. See also Lindqvist, "Avoidance of Sun Exposure as a Risk Factor for Major Causes of Death: A Competing Risk Analysis of the Melanoma in Southern Sweden Cohort," *Journal of Internal Medicine*, 2016.

103 *Lindqvist and his coauthors made an important point:* Lindqvist, "The Relationship Between Sun Exposure and All-Cause Mortality," *PPS*, 2016.

104 *an accompanying commentary from Penn State's:* Jablonski, "Is There a Golden Mean for Sun Exposure?," *Journal of Internal Medicine*, 2014.

CHAPTER 7 RELAX

106 *the sun had actually dropped a big hint:* For Furchgott's account of this history, see Furchgott, "The Discovery of Endothelium-Derived Relaxing Factor and Its Importance in the Identification of Nitric Oxide," *JAMA*, 1996.

107 *He called the effect photorelaxation:* Furchgott, "The Photoactivated Relaxation of Smooth Muscle of Rabbit Aorta," *Journal of General Physiology*, 1961.

109 *Weller knew that blood pressure was higher:* Gemmell, "Seasonal Variation in Mortality in Scotland," *International Journal of Epidemiology*, 2000. For an overview on the seasonal variations in disease, see Marti-Soler, "Seasonal Variation of Overall and Cardiovascular Mortality: A Study in 19 Countries from Different Geographic Locations," *PLOS One*, 2014.

110 *"Sent it off to the* Journal of Investigative Dermatology*":* Weller, "Nitric Oxide Is Generated on the Skin Surface by Reduction of Sweat Nitrate," *JID*, 1996.

111 *they swapped research notes:* Weller's German colleague Christoph Suschek was also responsible for some of the key discoveries. See Paunel, "Enzyme-Independent Nitric Oxide Formation During UVA Challenge of Human Skin: Characterization, Molecular Sources, and Mechanisms," *Free Radical Biology and Medicine*, 2005; Oplander, "Whole Body UVA Irradiation Lowers Systemic Blood Pressure by Release of Nitric Oxide from Intracutaneous Photolabile Nitric Oxide Derivates," *Circulation Research*, 2009; and Oplander, "New Aspects of Nitrite Homeostasis in Human Skin," *JID*, 2009.

111 *He had coauthored a paper showing that ultraviolet light:* Dejam, "Thiols Enhance NO Formation from Nitrate Photolysis," *Free Radical Biology and Medicine*, 2003.

113 *Weller's first appeared in the* Journal of Investigative Dermatology*:* Mowbray, "Enzyme-Independent NO Stores in Human Skin: Quantification and Influence of UV Radiation," *JID*, 2009.

113 *His second paper,* written with Martin Feelisch *in 2010:* Feelisch, "Is Sunlight Good for Our Heart?," *European Heart Journal*, 2010.

114 *as Weller pointed out in his paper:* Liu, "UVA Irradiation of Human Skin Vasodilates Arterial Vasculature and Lowers Blood Pressure Independently of Nitric Oxide Synthase," *JID*, 2014.

114 *Weller checked a three-year record of the blood pressure:* Weller, "Does Incident Solar Ultraviolet Radiation Lower Blood Pressure?," *Journal of the American Heart Association*, 2020.

114 *Since then the effect has repeatedly been confirmed:* See Hazell, "Post-Exposure Persistence of Nitric Oxide Upregulation in Skin Cells Irradiated by UV-A," *Scientific Reports*, 2022; and Hazell, "Low-Dose Daylight Exposure Induces Nitric Oxide Release and Maintains Cell Viability in Vitro," *Scientific Reports*, 2023.

CHAPTER 8 ALL CAUSE

116 *researchers were shining UV light on mice:* See the work of Shelley Gorman for papers on why sunlight seems to rev up metabolism and protect against obesity and cardiovascular disease: Gorman, "Ultraviolet Radiation, Vitamin D and the Development of Obesity, Metabolic Syndrome and Type-2 Diabetes," *PPS*, 2016; Gorman, "Sun Exposure: An Environmental Preventer of Metabolic Dysfunction?," *Current Opinion in Endocrine and Metabolic Research*, 2020; Geldenhuys, "Ultraviolet Radiation Suppresses Obesity and Symptoms of Metabolic Syndrome Independently of Vitamin D in Mice Fed a High-Fat Diet," *Diabetes*, 2014; and Fleury, "Sun Exposure and Its Effects on Human Health: Mechanisms Through Which Sun Exposure Could Reduce the Risk of Developing Obesity and Cardiometabolic Dysfunction," *International Journal of Environmental Research and Public Health*, 2016.

116 *In the UK and the Netherlands, a study:* Noordam, "Associations of Outdoor Temperature, Bright Sunlight, and Cardiometabolic Traits in Two European Population-Based Cohorts," *Journal of Clinical Endocrinology and Metabolism*, 2019.

117 *They concluded with the now-immortal words:* Lindqvist, "Avoidance of Sun Exposure as a Risk Factor for Major Causes of Death: A Competing Risk Analysis of the Melanoma in Southern Sweden Cohort," *Journal of Internal Medicine*, 2016.

117 *Its president fired off a letter to the* Journal*:* Torres, "Response to 'Avoidance of Sun Exposure as a Risk Factor for Major Causes of Death: A Competing Risk Analysis of the Melanoma in Southern Sweden Cohort,'" *Journal of Internal Medicine*, 2016.

117 *Lindqvist and Olsson fired back:* Lindqvist, "Answer to IM-16-0459," *Journal of Internal Medicine*, 2016. Lindqvist also summarized his work in this area in a 2018 review: Lindqvist, "The Winding Path Toward an Inverse Relationship Between Sun Exposure and All-Cause Mortality," *Anticancer Research*, 2018.

118 *Dibben and Weller found:* Stevenson, "Higher Ultraviolet Light Exposure Is Associated with Lower Mortality: An Analysis of Data from the UK Biobank Cohort Study," *Health and Place*, 2024.

119 *Weller got curious about how many people:* Best source for this is Weller's 2025 lecture at the Guy Foundation, available at the foundation's website: Weller, "The Health Effects of Sunlight, UV, and Blue Light," Guy Foundation, 2025.

120 *A 2020 Brazilian study fitted 103 subjects:* Benedito-Silva, "Association Between Light Exposure and Metabolic Syndrome in a Rural Brazilian Town," *PLOS One*, 2020. See also a later study on light at night and metabolic syndrome: Hu, "Outdoor Light at Night Is a Modifiable Environmental

Factor for Metabolic Syndrome: The 33 Communities Chinese Health Study," *Science of the Total Environment*, 2024.

120 *the conclusions of the giant 2024 UK Biobank study:* Windred, "Brighter Nights and Darker Days Predict Higher Mortality Risk," *PNAS*, 2024.

120 *Weller also found that high-sun Brits:* Weller, "Health Effects of Sunlight."

CHAPTER 9 FRIENDLY FIRE

123 *the steepness of the MS latitude gradient:* See Simpson, "Latitude Continues to Be Significantly Associated with the Prevalence of Multiple Sclerosis: An Updated Meta-Analysis," *Journal of Neurological and Neurosurgical Psychiatry*, 2019; and Sabel, "The Latitude Gradient for Multiple Sclerosis Prevalence Is Established in the Early Life Course," *Brain*, 2021. The first recorded note of it was Davenport, "Multiple Sclerosis," *Archives of Neurology and Psychiatry*, 1922.

123 *a 1960 study of veterans:* Acheson, "Some Comments on the Relationship of the Distribution of Multiple Sclerosis to Latitude, Solar Radiation, and Other Variables," *Acta Psychiatrica Scandinavica*, 1960.

124 *Some of the best data come from Lucas's home country:* Hammond, "The Epidemiology of Multiple Sclerosis in Three Australian Cities: Perth, Newcastle and Hobart," *Brain*, 1981. Also see Langer-Gould, "MS Sunshine Study: Sun Exposure but Not Vitamin D Is Associated with Multiple Sclerosis Risk in Blacks and Hispanics," *Nutrients*, 2018; and Gallagher, "Lifetime Exposure to Ultraviolet Radiation and the Risk of Multiple Sclerosis in the US Radiologic Technologists Cohort Study," *Multiple Sclerosis Journal*, 2019.

124 *pretty much everywhere she looked:* Lucas, "Sun Exposure and Vitamin D Are Independent Risk Factors for CNS Demyelination," *Neurology*, 2011; Lucas, "Ultraviolet Radiation, Vitamin D and Multiple Sclerosis," *Neurodegenerative Disease Management*, 2015; Simpson, "Sun Exposure Across the Life Course Significantly Modulates Early Multiple Sclerosis Clinical Course," *Frontiers in Neurology*, 2018; Wallin, "The Prevalence of MS in the United States," *Neurology*, 2019; and Hedstrom, "Low Sun Exposure Increases Multiple Sclerosis Risk Both Directly and Indirectly," *Journal of Neurology*, 2020.

124 *just one-third the rate of MS:* Van der mei, "Past Exposure to Sun, Skin Phenotype, and Risk of Multiple Sclerosis: Case-Control Study," *BMJ*, 2003.

125 *Incidence rates correspond to month of birth:* Staples, "Low Maternal Exposure to Ultraviolet Radiation in Pregnancy, Month of Birth, and Risk of Multiple Sclerosis in Offspring: Longitudinal Analysis," *BMJ*, 2010.

125 *Kids who spend less than thirty minutes a day:* Sebastian, "Association Between Time Spent Outdoors and Risk of Multiple Sclerosis," *Neurology*, 2022.

125 *a handful of small pilot studies:* See Breuer, "UVB Light Attenuates the Systemic Immune Response in CNS Autoimmunity," *Annals of Neurology*, 2014; and Ostkamp, "Sunlight Exposure Exerts Immunomodulatory Effects to Reduce Multiple Sclerosis Severity," *PNAS*, 2021.

125 *The most interesting study was by Prue Hart:* A remarkable number of papers have come out of Hart's clinical trial, including Hart, "A Randomised, Controlled Clinical Trial of Narrowband UVB Phototherapy for Clinically Isolated Syndrome: The PhoCIS Study," *Multiple Sclerosis Journal*, 2018; Trend, "Narrowband UVB Phototherapy Reduces TNF Production by B-Cell Subsets Stimulated via TLR7 from Individuals with Early Multiple Sclerosis," *Clinical and Translational Immunology*, 2020; and Hart, "Proteomics Confirms Immune Stabilizing Effects of Narrowband UVB Treatment in Patients with Clinically Isolated Syndrome and Multiple Sclerosis," *Multiple Sclerosis and Related Disorders*, 2025.

125 *"Have we reached the threshold to say":* In 2018, Lucas published a fascinating paper comparing the quality of evidence proving smoking causes lung cancer and the evidence connecting sun exposure and MS: Lucas, "On the Nature of Evidence and 'Proving' Causality: Smoking and Lung Cancer vs. Sun Exposure, Vitamin D and Multiple Sclerosis," *International Journal of Environmental Research*, 2018.

126 *The effect shows up in related diseases, too:* See Miller, "Are Low Sun Exposure and/or Vitamin D Risk Factors for Type 1 Diabetes?," *PPS*, 2017; Kahn, "Association of Type 1 Diabetes with Month of Birth Among U.S. Youth," *Diabetes Care*, 2009; Bora, "The Role of UVR and Vitamin D on T Cells and Inflammatory Bowel Disease," *PPS*, 2016; and Miller, "Higher Ultraviolet Radiation During Early Life Is Associated with Lower Risk of Childhood Type 1 Diabetes Among Boys," *Scientific Reports*, 2021.

126 *when a team of Australian researchers attached:* Reuter, "Direct Infant UV Light Exposure Is Associated with Eczema and Immune Development," *Journal of Allergy and Clinical Immunology*, 2019.

126 *which was also true for the MS patients:* Hart, "Proteomics Confirms Immune Stabilizing."

127 *any type of stressor that threatens to throw:* For some good overviews on inflammation and homeostasis, see Hotamisligil, "Inflammation, Metaflammation and Immunometabolic Disorders," *Nature*, 2017; and Chovatiya, "Stress, Inflammation, and Defense of Homeostasis," *Molecular Cell*, 2014.

127 *Rates of autoimmune diseases have been rising:* Miller, "The Increasing Prevalence of Autoimmunity and Autoimmune Diseases: An Urgent Call to

Action for Improved Understanding, Diagnosis, Treatment and Prevention," *Current Opinion in Immunology*, 2023.

128 *The field even has its own name:* photoimmunology*:* For a history of the photoimmunology field, see Ullrich, "The Immunologic Revolution: Photoimmunology," *JID*, 2012.

128 *In 1974, Margaret Kripke:* Kripke, "Antigenicity of Murine Skin Tumors Induced by Ultraviolet Light," *Journal of the National Cancer Institute*, 1974.

128 *"That was the key!":* Kripke's account of her work can be found in Kripke, "Reflections on the Field of Photoimmunology," *JID*, 2013.

129 *In 2025, a team at the University of Lyon:* Urocanic acid, or UCA, is produced in the skin in a trans form that flips to its cis isomer when hit by UV light. The cis form seems to be good food for the skin microbiome. See Patra, "Urocanase-Positive Skin-Resident Bacteria Metabolize Cis-Urocanic Acid and In Turn Reduce the Immunosuppressive Properties of UVR," *JID*, 2025. See also the commentary in the same issue of *JID*: Camera, "A Microbial Shield: Interactions Between UVR, Cis-Urocanic Acid, and the Skin Microbiome," *JID*, 2025.

129 *urocanic acid (UCA), one of the main molecules:* For an overview of the many roles of UCA in health, see Hart, "The Multiple Roles of Urocanic Acid in Health and Disease," *JID*, 2021.

129 *Later studies found that broad-spectrum sunscreens:* Moyal, "Broad-Spectrum Sunscreens Provide Better Protection from Solar Ultraviolet-Simulated Radiation and Natural Sunlight-Induced Immunosuppression in Human Beings," *JAAD*, 2008.

130 *Other researchers have learned that the immunosuppression:* See Schwarz, "25 Years of UV-Induced Immunosuppression Mediated by T Cells—from Disregarded T Suppressor Cells to Highly Respected Regulatory T Cells," *PPS*, 2008; Schwarz, "The Dark and the Sunny Sides of UVR-Induced Immunosuppression: Photoimmunology Revisited," *JID*, 2010; and Tse, "Exposure to Solar Ultraviolet Radiation Establishes a Novel Immune Suppressive Lipidome in Skin-Draining Lymph Nodes," *Frontiers in Immunology*, 2023.

130 *In such an environment, the immune system has to learn:* For some good overviews on the ways light helps to train the immune system to be tolerant, see Hart, "Exposure to Ultraviolet Radiation in the Modulation of Human Diseases," *Annual Review of Pathology: Mechanisms of Disease*, 2019; Bernard, "Photoimmunology: How Ultraviolet Radiation Affects the Immune System," *Nature Reviews Immunology*, 2019; and Bernard, "Ultraviolet Radiation Damages Self Noncoding RNA and Is Detected by TLR3," *Nature Medicine*, 2012.

130 *Kids introduced to small amounts of peanut:* Gabryszewski, "Guidelines for Early Food Introduction and Patterns of Food Allergy," *Pediatrics*, 2025.

131 *And to be tolerant:* See Martins, "Disease Tolerance as an Inherent Component of Immunity," *Annual Review of Immunology*, 2019.

131 *Kripke reflected years later:* Kripke, "Ultraviolet Radiation and Immunology: Something New Under the Sun—Presidential Address," *Cancer Research*, 1994.

131 *One fascinating bit of confirmation:* Lembo, "Polymorphic Light Eruption and Skin Cancer Prevalence: Is One Protective Against the Other?" *BJD*, 2008.

132 *A 2021 study of nearly a million psoriasis patients:* Hong, "The Potential Impact of Systemic Anti-Inflammatory Therapies in Psoriasis on Major Adverse Cardiovascular Events: A Korean Nationwide Cohort Study," *Scientific Reports*, 2021.

132 *another 2021 study of more than twelve thousand:* Bae, "Both Cardiovascular and Cerebrovascular Events Are Decreased Following Long-Term NB-UVB Phototherapy in Patients with Vitiligo: A Propensity-Score Matching Analysis," *Journal of the European Academy of Dermatology and Venereology*, 2021.

133 *Cardiovascular disease is often caused:* Sasaki, "UVB Exposure Prevents Atherosclerosis by Regulating Immunoinflammatory Responses," *Arteriosclerosis, Thrombosis, and Vascular Biology*, 2017.

133 *Alzheimer's is connected to low-grade:* See Weaver, "Alzheimer's Disease as an Innate Autoimmune Disease (AD2): A New Molecular Paradigm," *Alzheimer's and Dementia*, 2023; and Meier-Stephenson, "Alzheimer's Disease as an Autoimmune Disorder of Innate Immunity Endogenously Modulated by Tryptophan Metabolites," *Translational Research and Clinical Interventions*, 2022.

133 *"Inflammation: The Cause of All Diseases":* Chavda, "Inflammation: The Cause of All Diseases," *Cells*, 2024.

133 *the problem gets worse as we age:* Furman, "Chronic Inflammation in the Etiology of Disease Across the Life Span," *Nature Medicine*, 2019.

134 *So, counterintuitively, exercise is anti-inflammatory:* Here's a sheaf of articles on exercise's hormetic (anti-inflammatory) effects: Peake, "Modulating Exercise-Induced Hormesis: Does Less Equal More?," *Journal of Applied Physiology*, 2015; Luo, "The Anti-Inflammatory Effects of Exercise on Autoimmune Diseases: A 20-year Systematic Review," *Journal of Sports and Health Science*, 2024; Fiuza-Luces, "Exercise Benefits in Cardiovascular Disease: Beyond Attenuation of Traditional Risk Factors," *Nature Reviews*, 2018; Tryfidou, "DNA Damage Following Acute Aerobic Exercise: A Systematic Review and Meta-Analysis," *Sports Medicine*, 2020; Pesheva, "Research

Shows Working Out Gets Inflammation-Fighting T Cells Moving," *Harvard Gazette*, 2023. Also see Daniel Lieberman's book *Exercised.*

135 *Part of the reason plant-rich diets:* Martucci, "Mediterranean Diet and Inflammaging Within the Hormesis Paradigm," *Nutrition Reviews*, 2017.

CHAPTER 10 IN THE RED ZONE

138 *The main molecules in the body:* Powner, "Light Stimulation of Mitochondria Reduces Blood Glucose Levels," *Journal of Biophotonics*, 2024.

140 *The field traces its origins:* For a history of PBM, see Mester, "The History of Photobiomodulation: Endre Mester (1903–1984)," *Photomedicine and Laser Surgery*, 2017; and Hamblin, "Photobiomodulation or Low-level Laser Therapy," *Journal of Biophotonics*, 2016.

141 *The whole field might have remained fringe:* "NASA Research Illuminates Medical Uses of Light," NASA Technology Transfer Program, 2022.

142 *"It is essential for us to continue":* "Pushing Photodermatology into Practice," *AAD Meeting News*, 2024.

142 *Infrared light also seems to improve the function of lung cells:* Özdemir, "Effect of Photobiomodulation Therapy on Surfactant Production Increase in Human Lung Epithelial Alveolar Cells," *PPS*, 2025.

144 *Zimmerman found that the same thing happens:* Zimmerman, "Melatonin and the Optics of the Human Body," *Melatonin Research*, 2019.

145 *In 2025, Fosbury teamed up with Glen Jeffery:* Jeffery, "Longer Wavelengths in Sunlight Pass Through the Human Body and Have a Systemic Impact Which Improves Vision," *Scientific Reports*, 2025.

CHAPTER 11 TICK TOCK

152 *shift workers suffer alarmingly high rates:* For some depressing studies on shift work and health, see Streng, "Night Shift Work Characteristics Are Associated with Several Elevated Metabolic Risk Factors and Immune Cell Counts in a Cross-Sectional Study," *Scientific Reports*, 2022; and Caruso, "Negative Impacts of Shiftwork and Long Work Hours," *Rehabilitation Nursing Journal*, 2013. See also the books *Life Time* by Russell Foster (Yale, 2022) and *Why We Sleep* by Matthew Walker (Scribner, 2017).

153 *the WHO now classifies shift work:* "IARC Monographs Volume 124: Night Shift Work," International Agency for Research on Cancer, 2020.

153 *volunteers who spent a night of sleep in 100 lux:* Mason, "Light Exposure During Sleep Impairs Cardiometabolic Function," *PNAS*, 2022.

153 *Even scarier data was reported in 2025:* "Exposure to More Artificial Light at Night May Raise Heart Disease Risk," American Heart Association newsroom, 2025.

154 *The scariest data of all come from that big UK Biobank study:* Windred, "Brighter Nights and Darker Days Predict Higher Mortality Risk," *PNAS*, 2024. Intriguingly, that same research group also found that sleep regularity—consistency of bedtimes and rising times—was more important for reducing mortality risk than sleep duration: Windred, "Sleep Regularity Is a Stronger Predictor of Mortality Risk than Sleep Duration: A Prospective Cohort Study," *Sleep*, 2024.

156 *is what triggers the need to sleep:* This was a revelatory finding by the University of Oxford researchers Gero Miesenböck and Raffaele Sarnataro, and it made a big splash in *Nature* in 2025. They found that when mitochondria go about their normal business of processing energy for our cells, a certain number of electrons invariably leak out and form reactive oxygen species, which are free radicals that damage other molecules around them. Certain neurons in the brain have structures for measuring the buildup of this "electrical stress," and when it crosses a threshold they start signaling the brain to go to sleep. When enough of them start sending the sleep signal, we can't resist it anymore. See: Sarnataro, "Mitochondrial Origins of the Pressure to Sleep," *Nature*, 2025.

157 *One classic study by Harvard's Charles Czeisler:* Chang, "Evening Use of Light-Emitting eReaders Negatively Affects Sleep, Circadian Timing, and Next-Morning Alertness," *PNAS*, 2014. See also Chinoy, "Unrestricted Evening Use of Light-Emitting Tablet Computers Delays Self-Selected Bedtime and Disrupts Circadian Timing and Alertness," *Physiology Reports*, 2018.

158 *the average amount of ambient light it took to suppress:* Hartstein, "High Sensitivity of Melatonin Suppression Response to Evening Light in Preschool-Aged Children," *Journal of Pineal Research*, 2023; and Hartstein, "The Circadian Response to Evening Light Spectra in Early Childhood: Preliminary Insights," *Journal of Biological Rhythms*, 2025.

CHAPTER 12 DARKNESS ON THE EDGE OF TOWN

164 *Her specialty is "mind after midnight":* Tubbs, "The Mind After Midnight: Nocturnal Wakefulness, Behavioral Dysregulation, and Psychopathology," *Frontiers in Network Physiology*, 2022.

165 *"We found that kids are just absurdly sensitive":* See Hartstein, "High Sensitivity of Melatonin Suppression Response to Evening Light in Preschool-Aged

Children," *Journal of Pineal Research*, 2022; Hartstein, "Evening Light Intensity and Phase Delay of the Circadian Clock in Early Childhood," *Journal of Biological Rhythms*, 2023; and Hartstein, "The Circadian Response to Evening Light Spectra in Early Childhood: Preliminary Insights," *Journal of Biological Rhythms*, 2025.

167 *during the day, it helps suppress melatonin:* Although melatonin gets the lion's share of the attention when it comes to sleep and circadian issues, I suspect cortisol is equally important and is going to get more attention in the future. (And it shouldn't be confused with adrenaline, the other hormone produced by the adrenal gland. Adrenaline is the fight-or-flight, kick-in-the-pants hormone that supercharges your immediate response. It can save your life in the moment, but it's a shock to the system and you don't want it lasting more than a few minutes. Cortisol comes on more slowly, sharpening focus and mobilizing long-term energy reserves to deal with a less scary but more ongoing stressor—such as the SAT, or just the slings and arrows of daily life.) For a review of how different wavelengths of light affect cortisol, see Robertson-Dixon, "The Influence of Light Wavelength on Human HPA Axis Rhythms: A Systematic Review," *Life*, 2023. And for an experiment where blue-enriched light made students less drowsy and more alert, see Choi, "Awakening Effects of Blue-Enriched Morning Light Exposure on University Students' Physiological and Subjective Responses," *Scientific Reports*, 2019.

CHAPTER 13 THE LIGHT FANTASTIC

173 *The southeast-facing rooms received twice as much:* Choi, "Impacts of Indoor Daylight Environments on Patient Average Length of Stay (ALOS) in a Healthcare Facility," *Building and Environment*, 2012.

173 *Another Korean study of eighty thousand hospitalized patients:* Park, "The Effects of Natural Daylight on Length of Hospital Stay," *Environmental Health Insights*, 2018.

174 *Researchers in Copenhagen measured the light:* Gbyl, "Depressed Patients Hospitalized in Southeast-Facing Rooms Are Discharged Earlier than Patients in Northwest-Facing Rooms," *Neuropsychobiology*, 2017.

175 *the Harvard neuroscientist Rami Burstein discovered:* For research on green light, see Noseda, "A Neural Mechanism for Exacerbation of Headache by Light," *Nature Neuroscience*, 2010; Noseda, "Migraine Photophobia Originating in Cone-Driven Retinal Pathways," *Brain*, 2016; Ibrahim, "Long-Lasting Antinociceptive Effects of Green Light in Acute and Chronic Pain in Rats, *Pain*, 2017; and Lipton, "Narrow Band Green Light Effects on Headache,

Photophobia, Sleep, and Anxiety Among Migraine Patients: An Open-label Study Conducted Online Using Daily Headache Diary," *Frontiers in Neurology*, 2023.

175 *"Almost unanimously, they describe green light":* For Burstein's interview, see "Understanding Migraine and Light Sensitivity," *Association of Migraine Disorders*, 2020.

175 *Mohab Ibrahim now uses green light:* Martin, "Evaluation of Green Light Exposure on Headache Frequency and Quality of Life in Migraine Patients: A Preliminary One-Way Cross-Over Clinical Trial," *Cephalalgia*, 2021. For an interview with Ibrahim, see Kreier, "Going Green (Light) for Migraine and Pain: An Interview with Mohab Ibrahim," *Migraine Science Collaborative*, 2024.

176 *A few years ago, the light pioneer George Brainard:* Brainard, "Solid-State Lighting for the International Space Station: Tests of Visual Performance and Melatonin Regulation," *Acta Astronautica*, 2013; Brainard, "Short Wavelength Enrichment of Polychromatic Light Enhances Human Melatonin Suppression Potency," *Journal of Pineal Research*, 2015; "Lighting in a Bottle," *NASA Spinoff*, 2022; and "The Light Fantastic," *Thomas Jefferson University Alumni Magazine*, 2023.

176 *She pointed to some recent studies by David Samson:* Samson, "Are Humans Facing a Sleep Epidemic or Enlightenment? Large-Scale, Industrial Societies Exhibit Long, Efficient Sleep yet Weak Circadian Function," *Proceedings of the Royal Society B*, 2024; Samson, "Chronotype Variation Drives Night-Time Sentinel-Like Behaviour in Hunter-Gatherers," *Proceedings of the Royal Society B*, 2017; Samson, "Sleep Intensity and the Evolution of Human Cognition," *Evolutionary Anthropology*, 2015; Samson, "The Evolution of Human Sleep: Technological and Cultural Innovation Associated with Sleep-Wake Regulation Among Hadza Hunter-Gatherers," *Journal of Human Evolution*, 2017; and Samson, "Hadza Sleep Biology: Evidence for Flexible Sleep-Wake Patterns in Hunter-Gatherers," *American Journal of Physical Anthropology*, 2017. See also Yetish, "Natural Sleep and Its Seasonal Variations in Three Pre-Industrial Societies," *Current Biology*, 2015; and de la Iglesia, "Access to Electric Light Is Associated with Shorter Sleep Duration in a Traditionally Hunter-Gatherer Community," *Journal of Biological Rhythms*, 2015.

177 *That was also the gist of a recent study:* Wright, "Entrainment of the Human Circadian Clock to the Natural Light-Dark Cycle," *Current Biology*, 2013.

178 *This shows up in the cancer rates:* Gu, "Longitude Position in a Timezone and Cancer Risk in the United States," *Cancer Epidemiology, Biomarkers, and Prevention*, 2017.

178 *Another study found that people on the wrong end:* Giuntella, "Sunset Time and the Economic Effects of Social Jetlag: Evidence from US Time Zone Borders," *Journal of Health Economics*, 2019.

CHAPTER 14 TAKING THE CURE

180 *To find out, a Parisian research team first exposed:* Menezes, "Non-Coherent Near Infrared Radiation Protects Normal Human Dermal Fibroblasts from Solar Ultraviolet Toxicity," *JID*, 1998. See also Frank, "Infrared Radiation Induces the p53 Signaling Pathway: Role in Infrared Prevention of Ultraviolet B Toxicity," *Experimental Dermatology*, 2006.

181 *A related experiment by the McGill University dermatologist:* Barolet, "Infrared and Skin: Friend or Foe?," *Journal of Photochemistry and Photobiology*, 2016; and Barolet, "LED Photoprevention: Reduced MED Response Following Multiple LED Exposures," *Lasers in Surgery and Medicine*, 2008.

181 *Barolet has also found:* Barolet, "Light-Induced Nitric Oxide Release in the Skin Beyond UVA and Blue Light: Red & Near-Infrared Wavelengths," *Nitric Oxide*, 2021; and Barolet, "Differential Nitric Oxide Responses in Primary Cultured Keratinocytes and Fibroblasts to Visible and Near-Infrared Light," *Antioxidants*, 2024.

183 *It's more of an observatory:* For a trippy overview of how skin senses light, see Monteiro de Assis, "How Does the Skin Sense Sun Light? An Integrative View of Light-Sensing Molecules," *Journal of Photochemistry and Photobiology C*, 2021.

184 *That's become even clearer:* Shraim, "Genome-Wide Gene-Environment Interaction Study Uncovers 162 Vitamin D Status Variants Using a Precise Ambient UVB Measure," *Nature Communications*, 2025. See also Andrzej Slominski's recent paper postulating the same hypothesis: Slominski, "Is Vitamin D Signaling Regulated by and Does It Regulate Circadian Rhythms?," *FASEB Journal*, 2025.

184 *the skin stocked itself with antioxidants:* One extremely important antioxidant in the body that often gets overlooked is . . . melatonin. Although most research focuses on its role in the brain as a herald of the news that darkness has fallen, new research has discovered that it is produced by virtually all cells in the body as an antioxidant. We didn't know this because it rarely leaves the cell interiors, so we don't detect it in the blood. But inside those cells, it is mopping up free radicals all day long, and that might be its original role in life, millions of years before it was conveying information about day and night. Importantly, cells seem to crank up production in response

to sunlight—in preparation for the free-radical production to come. See Tan, "Melatonin: Both a Messenger of Darkness and a Participant in the Cellular Actions of Non-Visible Solar Radiation of Near Infrared Light," *Biology*, 2023.

186 *Japan's forest bathers seem to have figured:* The evidence that forest bathing can reduce blood pressure and stress and improve mental health is pretty solid. See Li, "Therapeutic Effects of Forest Bathing on Older Adult Patients with Essential Hypertension: Evidence from a Subtropical Evergreen Broad-Leaved Forest," *Frontiers in Public Health*, 2025.

CHAPTER 15 **RIGHT WITH LIGHT**

189 *Their new forty-four-page "Position Statement":* Neale, "Balancing the Risks and Benefits of Sun Exposure: A Revised Position Statement for Australian Adults," *Chronic Disease Prevention and Control*, 2023.

191 *"This is totally the wrong advice":* Baird-Murray, "Do You Actually Need to Wear SPF Every Day? Separating Fact from Fiction," *Vogue*, 2025.

191 *but experts believe they carry a small amount of risk:* To learn more about the risky chemicals in sunscreens, see Ruszkiewicz, "Neurotoxic Effect of Active Ingredients in Sunscreen Products, a Contemporary Review," *Toxicology Reports*, 2017; Downs, "Benzophenone Accumulates over Time from the Degradation of Octocrylene in Commercial Sunscreen Products," *Chemical Research in Toxicology*, 2021; and Wnuk, "Benzophenone-3, a Chemical UV-Filter in Cosmetics: Is It Really Safe for Children and Pregnant Women?," *Advances in Dermatology and Allergology*, 2022. Also see the sources for the rest of this chapter.

191 *It has flagged the most common active chemicals:* Matta, "Effect of Sunscreen Application on Plasma Concentration of Sunscreen Active Ingredients: A Randomized Clinical Trial," *JAMA*, 2020. For an overview, see "The Trouble with Sunscreen Ingredients," Environmental Working Group, 2025.

192 *in 2008, the CDC published a study:* See "CDC: Americans Carry Body Burden of Toxic Sunscreen Chemical," Environmental Working Group, 2008. See also DiNardo, "Dermatological and Environmental Toxicological Impact of the Sunscreen Ingredient Oxybenzone/Benzophenone-3," *Journal of Cosmetic Dermatology*, 2018.

192 *the FDA did its own studies:* Matta, "Effect of Sunscreen Application."

192 *"It is outrageous that over a decade ago":* LaMott, "Seven Sunscreen Chemicals Enter Bloodstream After One Use, FDA Says, but Don't Abandon Sun Protection," CNN, 2020.

192 *mounting evidence that the sunscreen running off:* For the evidence, see Downs, "Toxicopathological Effects of the Sunscreen UV Filter, Oxybenzone (Benzophenone-3), on Coral Planulae and Cultured Primary Cells and Its Environmental Contamination in Hawaii and the U.S. Virgin Islands," *Archives of Environmental Contamination and Toxicology*, 2015.

193 *A 2024 study at George Mason University:* Bloom, "Impact of Skin Care Products on Phthalates and Phthalate Replacements in Children: The ECHO-FGS," *Environmental Health Perspectives*, 2024.

193 *Even products that don't list PFAS:* Whitehead, "Fluorinated Compounds in North American Cosmetics," *Environmental Science and Technology Letters*, 2021; and Whitehead, "Directly Fluorinated Containers as a Source of Perfluoroalkyl Carboxylic Acids," *Environmental Science and Technology Letters*, 2023.

193 *Once again, toxicologists found:* Abraham, "Transdermal Absorption of 13C4-Perfluorooctanoic Acid (13C4-PFOA) from a Sunscreen in a Male Volunteer—What Could Be the Contribution of Cosmetics to the Internal Exposure of Perfluoroalkyl Substances (PFAS)?," *Environment International*, 2022; and Ragnarsdóttir, "Dermal Bioavailability of Perfluoroalkyl Substances Using In Vitro 3D Human Skin Equivalent Models," *Environment International*, 2024.

194 *Diffey has been calling for restraint:* Diffey, "Sun Protection: Have We Gone Too Far?," *BJD*, 1998; Diffey, "Is Daily Use of Sunscreens of Benefit in the U.K.?," *BJD*, 2002; and Diffey, "Sun Protection: False Beliefs and Misguided Advocacy," *BJD*, 2023.

194 *He later expanded that assessment:* Diffey, "When Should Sunscreen Be Applied: The Balance Between Health Benefit and Adverse Consequences to Humans and the Environment," *International Journal of Cosmetic Science*, 2023.

194 *in 2025, Australian investigators, led by Rachel Neale:* Tran, "The Effect of Daily Sunscreen Application on Vitamin D," *BJD*, 2025.

195 *there are lots of good goop-free strategies:* One Stanford study found that seeking shade and using clothing were much more effective than sunscreen at preventing sunburn. See Linos, "Hat, Shade, Long Sleeves, or Sunscreen? Rethinking US Sun Protection Messages Based on Their Relative Effectiveness," *Cancer Causes Control*, 2013.

198 *a study of 422,000 Scottish schoolchildren:* Hastie, "Antenatal Exposure to Solar Radiation and Learning Disabilities: Population Cohort Study of 422,512 Children," *Scientific Reports*, 2019.

INDEX

NOTE: Page numbers followed by "n" refer to footnotes. Page numbers in italics refer to "A Light Primer" insert.

ABOUT THE AUTHOR

Rowan Jacobsen's award-winning science writing has appeared in *Harper's*, *Outside*, *The Atlantic*, *Scientific American*, *Smithsonian*, *National Geographic*, *The New York Times*, *The Washington Post*, *MIT Technology Review*, *Businessweek*, and *The Best American Science & Nature Writing*. He has received multiple awards from the James Beard Foundation, the Society of American Travel Writers, the Overseas Press Club, and others. His ten books include *A Geography of Oysters*, *Fruitless Fall*, and *Wild Chocolate.* He has performed with *Pop-Up Magazine*, lectured at Harvard and Yale, and appeared on dozens of national radio and television programs. He has been an Alicia Patterson Foundation Fellow, writing about endangered diversity on the borderlands of India and Myanmar; a Knight Science Journalism Fellow at MIT, focusing on the promises and perils of synthetic biology; and a Media Fellow at the Nova Institute for Health, researching the science of light exposure. You can find him in Vermont—at least when the sun is shining.